Subversion, Conversion, D

**Infrastructures Series**
edited by Geoffrey Bowker and Paul N. Edwards

Paul N. Edwards, *A Vast Machine: Computer Models, Climate Data, and the Politics of Global Warming*

Lawrence M. Busch, *Standards: Recipes for Reality*

Lisa Gitelman, ed., *"Raw Data" Is an Oxymoron*

Finn Brunton, *Spam: A Shadow History of the Internet*

Nil Disco and Eda Kranakis, eds., *Cosmopolitan Commons: Sharing Resources and Risks across Borders*

Casper Bruun Jensen and Brit Ross Winthereik, *Monitoring Movements in Development Aid: Recursive Partnerships and Infrastructures*

James Leach and Lee Wilson, eds., *Subversion, Conversion, Development: Cross-Cultural Knowledge Encounter and the Politics of Design*

# Subversion, Conversion, Development

Cross-Cultural Knowledge Encounter and the Politics of Design

edited by James Leach and Lee Wilson

The MIT Press
Cambridge, Massachusetts
London, England

© 2014 Massachusetts Institute of Technology

All rights reserved. No part of this book may be reproduced in any form by any electronic or mechanical means (including photocopying, recording, or information storage and retrieval) without permission in writing from the publisher.

MIT Press books may be purchased at special quantity discounts for business or sales promotional use. For information, please email special_sales@mitpress.mit.edu.

This book was set in Stone by the MIT Press. Printed and bound in the United States of America.

Library of Congress Cataloging-in-Publication Data

Subversion, conversion, development : cross-cultural knowledge exchange and the politics of design / edited by James Leach and Lee Wilson.
    pages   cm.—(Infrastructures series)
Includes bibliographical references and index.
ISBN 978-0-262-02716-8 (hardcover : alk. paper)—ISBN 978-0-262-52583-1 (pbk. : alk. paper)
1. Information technology—Social aspects. 2. Technological innovation—Social aspects. 3. Community development—Case studies. 4. Internet and indigenous peoples. 5. Computers and civilization. I. Leach, James, 1969–, editor of compilation. II. Wilson, Lee, 1966–, editor of compilation.
HM851.S864  2014
303.48'34—dc23
2013035705

10  9  8  7  6  5  4  3  2  1

Dedicated to the memory of Catherine Mansell Gibson and Robert William Wilson.

# Contents

Acknowledgments  ix

1 Anthropology, Cross-Cultural Encounter, and the Politics of Design  1
James Leach and Lee Wilson

2 Liminal Futures: Poem for Islands at the Edge  19
Laura Watts

3 Freifunk: When Technology and Politics Assemble into Subversion  39
Gregers Petersen

4 Postcolonial Databasing? Subverting Old Appropriations, Developing New Associations  57
Helen Verran and Michael Christie

5 Sacred Books in a Digital Age: A Cross-Cultural Look from the Heart of Asia to South America  79
Hildegard Diemberger and Stephen Hugh-Jones

6 Redeploying Technologies: ICT for Greater Agency and Capacity for Political Engagement in the Kelabit Highlands  105
Poline Bala

7 Making the Invisible Visible: Designing Technology for Nonliterate Hunter-Gatherers  127
Jerome Lewis

8 Assembling Diverse Knowledges: Trails and Storied Spaces in Time  153
David Turnbull and Wade Chambers

9 Structuring the Social: Inside Software Design  183
Alan F. Blackwell

10  Design for X: Prediction and the Embeddedness (or Not) of Research in Technology Production   201
    Dawn Nafus

11  Engaging Interests   223
    Marilyn Strathern

12  Subversion, Conversion, Development: Imaginaries, Knowledge Forms, and the Uses of ICTs   231
    James Leach and Lee Wilson

Contributors   245
Index   247

# Acknowledgments

The editors gratefully acknowledge the contributions of participants at a workshop titled "Subversion, Conversion, Development: Public Interests in Information and Communication Technologies," convened by Robin Boast, James Leach, and Lee Wilson at the Centre for Research in the Arts, Social Sciences, and Humanities (CRASSH) at the University of Cambridge in late 2008. We especially thank Robin Boast for helping to organize and plan the meeting, and the Museum of Archaeology and Anthropology at the University of Cambridge for its financial assistance and support. We acknowledge Intel Labs and British Telecom for their contribution to funding the conference. CRASSH provided further financial support for the event through its conference award scheme. We wish also to thank Professor Mary Jacobus, the director of CRASSH, a source of both intellectual inspiration and support for the conference, and the tireless efforts of the staff at CRASSH, in particular Catherine Hurley and Anna Malinowska, in helping us to realize the event.

In giving shape and coherence to the manuscript, we have benefited from the guidance of Geof Bowker and from the suggestions of two anonymous reviewers for MIT Press. The importance of these contributions in shaping the final manuscript will be visible to them at least. We also thank Marguerite Avery at MIT Press for her support and encouragement.

# 1 Anthropology, Cross-Cultural Encounter, and the Politics of Design

James Leach and Lee Wilson

This collection presents a set of empirical cases and theoretical examinations that focus on alternative cultural encounters with and around information technologies (alternative, that is, to the dominant notions of media consumption among Western audiences). Some of the examples refer to media practices as forms of cultural resistance and subversion, some to DIY cultures and alternative models of technology production and appropriation. Then there are considerations of representation and political participation, of entwined practices of knowledge production and encoding, and of information capture and preservation. The contributors are interested in possibility, and they are interested in constraint. They reveal how certain epistemologies are built into mass-produced technological offerings and highlight the ways in which the hidden assumptions and politics behind design can be productively challenged or subverted in action (see also Lievrouw 2006; 2011, 3–5; Bolter and Grusin 1999). Our focus, then, is on instances where technologies and their design carry particular assumptions about social relations. These are particularly apparent where ubiquitous, neutral-seeming tools are revealed as carrying normative principles when they are used in unfamiliar or unexpected settings. Nowhere is this more apparent than when it comes to the treatment of "knowledge." Often approached as a kind of neutral good with universal reference, as something that can be abstracted from the context of its generation and made to carry value into other domains of action through new technologies without reference to the social relations of emergence, knowledge is central to our concern. Information and communications technologies tend to facilitate a process of flattening difference when it comes to knowledge and, in doing so, reveal their own organizing effect.

A "productionist" metaphysic (Weiner 2001) lies behind the contemporary visibility of knowledge in knowledge economies, made most apparent by its valuation under intellectual property regimes (see Rubio and Beart 2012). Attention to forms of collaborative work and uses of technology in this volume highlights the specific inflections that different cultures and communities give to the value of knowledge. These in turn challenge the effects of the media in which knowledge circulates, as well as the politics of design that facilitate a kind of appropriation or disembedding of value from social action.

In fact, the whole area of knowledge production in relation to media within knowledge economies needs opening up to further scrutiny (see Leach 2013). Today people tend to assume they know what is meant by "knowledge production," and devote huge effort to securing the correct conditions to realize this form of (economic) productivity and to circulate its products. But these processes tend to exclude many kinds of knowing (Harris 2007; Marchand 2010; Davis and Leach 2012) and undervalue the importance of exactly the kinds of emergent and relational "process-based" knowledge forms that anthropologists (for example) are familiar with. In contrast, this volume faces head-on the ways that different knowledge forms are part of particular social relations, and how that truth refigures or recasts the design or use of technology in specific instances.

Our consideration comes at a time when the area of knowledge is also being radically recast in the era of globalization and digitization. The concurrent emergence of free software as a model of production and collaboration (Ghosh 2005; Weber 2004), open research journals, online social networks (Benkler 2006), the digital preservation of heritage, and the multimedia presentation of art and performance (e.g., Morphy et al. 2005) relies on new modes for presenting and circulating things, practices, and understandings (Castells 1996; Brown 2003). These both use and challenge existing design parameters for media and communications technologies, and the assumptions behind them. The developments are made possible by the transformation of knowledge into kinds of information available for encoding and transmission (e.g., Gurstein 2000). But while being swept along in the current of technological change (E. Leach 1968), we should continually ask if all knowledge is simply amenable to this form. While some people who once were excluded from the circuits in which knowledge was produced are becoming integral parts of its production, others, such as indigenous knowledge holders, are often still excluded or participate on terms dictated elsewhere.

As new communities enter the field of knowledge production, we need to understand the kinds of knowledge they offer, and how and where media come to transform, facilitate, or undermine their validity. An example: a cross-cultural confusion about what to *do* with indigenous knowledge lies in conventional approaches to the production of agricultural staples or the production of artifacts in indigenous Papua New Guinean societies (Leach 2012). The confusion is based on assumptions about the separation of knowledge from relationships between people (Morphy 2007). That is, standard Euro-American conceptions of knowledge can be said to view it as representational of the world of nature (Shapin and Schaffer 1989; Latour 1993). Yet indigenous knowledge is often embedded in, as if it were in fact an aspect of, relations between persons and beings of different kinds (Morphy 1991; Strathern 2010). From horticultural practices that seem to rely on superstition and ritual to the embedded making of complex and valuable indigenous artworks in life cycle events, the entanglements of social and practical, natural and cultural, productive and decorative, have proved impossible to separate (see Strathern, this volume, on Lemonnier and Latour; Heckler 2009; Sillitoe 2009; Weiner 1995). Indigenous people themselves resist such separation (Takeshita 2001). This characteristic has tended to undermine the possibility of recognizing indigenous knowledge as commensurate with scientific or other forms of modern knowing (Crook 2007).

The challenges are particularly apparent when it comes to finding appropriate media forms to present knowledge, as the contributions gathered here make abundantly clear. Based in particular processes and effects, indigenous knowledge is increasingly under scrutiny for what it has to offer the contemporary world, yet has been difficult to access, translate, and specify on the same terms as scientific knowledge (Sillitoe et al. 2002). Similarly, while the contemporary arts are highly valued in Western societies, as are indigenous arts, a significant number of their practitioners are seeking to clarify and demonstrate the potential *knowledge* contributions their process-based creativity offers.

A series of questions follows from this contemporary issue, an issue that inextricably draws in design and epistemology. Perhaps most pertinent is the question of where we are going to find new resources to understand, value, and present such knowledge. It is a question about representation and the politics of design. How do we approach situated knowledge in a way acceptable to its source communities, in ways that avoid the old issues of appropriation, distortion, and confusion over its status while making value available and apparent? In attending to this question, this volume breaks new ground by focusing on specific cases where the forms of knowledge and the cross-cultural encounter are shaping technological use and development. In other words, we do not present critique alone but focus on positive examples of reshaping technologies by these encounters.

Ethnographic practice figures prominently in the contributions, many of which combine classical anthropological settings with new methodological and theoretical approaches. The studies complement research on the social shaping of technology as a matter of scientific development (Bijker and Law 1992; Ong and Collier 2004) and of everyday use (Horst and Miller 2006; Burrell 2009). Through ethnographic engagement, they show how models of collaborative work and of effective organization or action are realized or hindered by particular technologies.

As Alan Blackwell (this volume) notes, the practices of design and ethnography share a longer historical association than is commonly known. In the field of computer software design, Millen for example claims that ethnography is a useful methodology for gathering user requirements and iterative design (Millen 2000). Yet in the study of engineering or design work, prevailing misconceptions about what ethnographic practice might entail have led to the term being used as a proxy for fieldwork (Button 2000). Similarly, ethnography figures prominently in the development of the interdisciplinary field of science and technology studies, where the use of the term implies a variety of fieldwork practices and conventions of writing (Hess 2001, 243; see also Strathern 2004, 554). With this in mind, we outline here what we mean to imply when we talk of ethnography in the service of introducing the central concern of this volume: understanding the specificity of social relations and the constitutive role of those relations in the design and use of technology.

As a knowledge-making practice, ethnography is a means of reflecting on the world in new ways via close association with ideas, people, and things. As we understand the practice of ethnography, it usually entails gathering data to allow a description of events that makes explicit the presuppositions that underpin or inform action. The production of ethnographic accounts focuses on the ways in which the world is variously categorized, structured, and made sense of. The stuff of these accounts is multifarious and complex social relations. Ethnographic engagement with the subjects of study (or, indeed, the objectification of this encounter) is a personal and reflexive practice, a mode of engagement clearly demonstrated when ethnographers work with designers or engineers to make appropriate tools for use by a specific community. Ethnography is more than descriptive process. As a practice, it entails engagement, abstraction, reflection, and the production of second-order analysis (Bloor 1991) that results in the accuracy or adequacy of description (Runciman 1983, 224). It is an endeavor that requires sensitivity to, and awareness of, the ways in which knowledge is recognized in different social milieus and cultural contexts. We think that the production of ethnography is an inherently collaborative process (regardless of the degree to which these relationships are acknowledged in the accounts)

(J. Leach 2010). Indeed, as a mode of collaboration, ethnography can be used specifically as a means of critical engagement to generate accounts and objects with the specific intent of intervening in and transforming social contexts (see Wilson and Leitner 2009; Barab et al. 2004). And it is here, we would argue, that the value of ethnography as a process of critical engagement, and anthropology as a mode of critical reflection, is to be found. Advocates of critical approaches to technology design have for some time drawn on anthropology as a means to lay bare the implicit assumptions, values, and presuppositions that underpin both design and use (Agre 1997) and to render the familiar strange in critical design practice (Bell, Blythe, and Sengers 2005).

In preparing the volume, we found the notion of subversion to be particularly compelling. Subversion implies a sense of overturning or disrupting an established order. We employ it here in the sense of thwarting hegemony of form, of overturning structures of authority implicit in the process of design and manufacture. Subversion will also be apparent through the volume in the ways in which digital artifacts are generative of social relations whether these are novel or reconfigured social movements or communities. We have in mind, then, subversion of use, of purpose, and of corporate imaginings of the consumer purpose of technological artifacts, a "continuous design" or "redesign" of things (Tonkinwise 2005). The idea of conversion is similarly employed with the aim of bringing to light the reconfiguration and repurposing, even transmogrification, of digital artifacts. We explore approaches to ICT development that have adopted and enabled local modification and creativity.[1] This transformation and modification is not necessarily a radical process.

One has only to look at the rapid uptake and apparent importance of social media to see how very easily we can be seduced by the effectiveness of our technologies, particularly digital ICTs, because of their apparently incontrovertible transformative effects on political and social life. That seduction often results in a particular fetishization of the object: the power and effect of our technologies in achieving particular and interested outcomes generate an imaginary in which it is the technologies themselves that are powerful (Broers 2005; cf. J. Leach 2005). They are imagined as more than just tools for particular tasks and come to embody something fundamentally, progressively *human*. We seem liable to forget that the energy and excitement ICTs gather around them are socially generated specific energies and excitements. And in this context we turn to two examples of reuse and reworking that help introduce the contributions to come. These examples make concrete our intention to consider reconfigurability as an element of vital and creative social forms, and the specificity of social relations in the unfolding form of tools and things.

The contributors believe that it is important that we keep our focus on the generative aspects of using tools and technologies. That is how they figure in and partially configure particular social worlds. This is not to succumb to the same fetishization but to recognize the different possibilities that different forms of material and social engagement facilitate. We are interested in both the possibilities and the particularity of certain kinds of technologies. To illustrate this point more clearly, in the remainder of this introduction, we look at the design, production, and modification of specific objects. We return to the themes of cross-cultural knowledge encounter and the politics of design in the conclusion. The first example takes up things that may be unexpected in this context: dogs' teeth.

**The Social Purpose of Technologies**

Dogs' teeth are used as an exchange valuable on the Rai Coast of Papua New Guinea because of their importance as body decorations. How do they relate to the foregoing discussion on knowledge, locality, imaginaries, and a reconfigurable technology? We have a definite logic here that informs the volume as a whole. Some brief background is necessary for this logic to become apparent.

Dogs are vitally important in hunting for game on the Rai Coast. Game is both an everyday part of the diet and a ceremonial exchange item. Men hunt; women do not generally hunt for large game. Thus dogs represent aspects of a gendered social being. They are mobile and differentiated elements of a gendered intention and agency. One could quite easily think of the bows and arrows used by these people as technologies for hunting. Dogs are also part of hunting technology. Without the tool, the hunter is not a hunter (Gell 1998, 17, 198), and without being a hunter, an aspect of gender differentiation would be absent. If dogs are part of a technology of hunting, hunting is part of a technology of social relations in which persons come to have forms that complement one another, ensuring generation and social reproduction.

The world of Rai Coast people is populated by beings that are transformations of human forms. Hunting depends on spirits and ancestors that lie behind a man's head as he hunts, sitting at the nape of his neck. They are also present in landforms, streams, and animals in the forest. Many myths tell of people transforming back and forth from marsupials, or birds, and so on. So hunting is not just food procuring. It is certainly not (just) sport.

Hunting game is a dangerous, exciting, vital activity for Rai Coast men because it connects them to power, creation, mystery, and the spirit world. If hunting is an integral element in social reproduction, then hunters are at one and the same time getting food for their families and re-creating the world through their actions. The products of hunting (dried and smoked game) are pivotal to exchange and thus to kinship.

Dogs' teeth are removed from dogs and displayed at moments of social reproduction. So dogs' teeth become a partible item: a wealth item that carries value for the specific reason that they circulate the substance of life.

As an embodiment of male intention and agency, they bring with them knowledge of the actual work involved in specific previous hunts and exchanges that have produced particular people. Dogs' teeth, then, are of value as one outcome of productive work. If we think of them in this way, their use as display valuables amounts to a repurposing or, rather, a development of what is possible in the material (enamel teeth) through integral dialogue with, and thus development through, the social and conceptual lives of the people involved. But are dogs' teeth reconfigurable, open to reuse and repurposing?

It is not just men who wear dogs' teeth. *Dogs* are aspects of men's technology of hunting. But their teeth are used and displayed by both men and women. Women are not an audience for male display but the necessary other to a system of hunting as a male activity. Social reproduction is not seen as the province of one gender rather than the other. Upon her marriage, a bride moves to the land of her husband. She works there alongside her husband to produce the wealth that she will give to her kin.[2] In working to produce her substitute "body" to be given to her natal kin from these her adopted lands, a wife achieves two things. First, she is incorporated into the place she has moved to. The relations to others in which she grew are replaced by this work with new relations: to her new residence and her new kin, the spirits and people of these lands. Second, as a demonstration of this transformation, she achieves the right to be buried in this new place. When women wear dogs' teeth, it is the power of their place, of which they are fully part, that is displayed. When given as bride wealth, dogs' teeth may transmit an embodiment of a gendered agency from one village to another. But they are not inherently "male" or "female" valuables, because their value is generated in the wider productive work of both men and women.

Dogs' teeth are thus repurposed in the context of exchange. The value is already there, the significance established through use in the generation of a particular social form and the knowledge of the world necessary to it. The repurposing of the object is repurposing of that significance through reuse.

**Figure 1.1**
Sambong Porer displays a dogs' teeth back piece, 2006.

Anthropology, Cross-Cultural Encounter, and Design 9

**Figure 1.2**
Tariak Pulumamie wears a dogs' teeth headband and necklace during a marriage exchange, 2006.

We can agree, then, that dogs' teeth are reconfigured (rather prosaically) as display items redesigned every time individual teeth are put into new combinations. More deeply, their display is a repurposing of male power. That is, dogs' teeth that reference hunting are a gendered principle or effect in movable and partible form, not a fixed, gendered identity but available as power to persons of all kinds. It is in their very reconfigurability that they are elegant and valuable. And that is a matter not of their inherent practicality but of the social form they participate in constituting.

So the chapters in this volume address the way social relations that depend on and produce material things are reconfigured, facilitated, or obviated by the objects and technologies that are integral to their particular, historically constituted social forms.

A series of expectations, and therefore values, is built into any technology. This is both apparent in the tools and objects referred to by the authors here (databases, GPS receivers, books, computers) and also hidden, obscured by the ease of certain kinds of communication and therefore relation made possible with them. As many of the chapters elucidate, values and ways of understanding are built into modes of storage and communication. What technology producers have privileged is the communication aspect, and that comes down to information transfer. But information transfer is only a small part of communication. We might know this, but as more and more of our time and attention are taken up by using ICTs to make and sustain our social relations, as they give more form to our lives, other aspects *do* become obscured. Technology is not neutral, in other words; it is not value-free possibility. ICTs are not an inert medium to amplify everyone's social worlds and values. Dogs' teeth equally are not an inert medium though which Rai Coast people's social world is produced and reproduced. Their forms, and the forms of their transformations, are shaping aspects of a particular social and physical existence. In this, they resemble many examples of use, development, and repurposing in this volume. The particular configurations of power and an intricate specificity to social relations are made apparent in the motile life of dogs' teeth. But we could also point to examples that seem more familiar. The use and adaptation of urban buildings and space are a case in point.

## Designing for Relationships

Levittown, a planned community of some 17,500 houses built on Long Island, New York, after World War II, has been heralded as the dawn of mass-produced suburbia in America (Wagner and Wagner 2010, 7).

Characterized by conformity and hegemony of form, Levittown was a community defined by restrictive housing covenants and racial segregation.[3] In many ways, Levittown might be seen as the epitome of the racism and normative gender relations typical of the planned communities that sprang up around the peripheries of cities in postwar America. This odious aspect of the history of Levittown acknowledged, it is the dynamic process of adaptation and improvisation that took place that we wish to draw attention to.

Intending originally to provide low-cost housing for returning war veterans and their families, William Levitt employed construction techniques modeled on the assembly line system used by Henry Ford for his town. Cheap building materials, modularized construction, and a highly specialized workforce enabled Levitt's standardized vision of utopia to be produced at an accelerated rate. With mortgages underwritten by the Federal Housing Administration and the GI Bill, Levittown signaled the realization of the American dream of home ownership as an affordable reality (Kelly 1988). Veterans and their families flocked to the new development; in one day alone, 650 homes sold in just five hours (Ferrer and Navarra 1993, 17). While the demand for housing in Levittown ran high, many were critical of the new community. The architectural critic Lewis Mumford described the community as "an incipient slum" and wrote scathingly about its "uniform, unidentifiable houses ... inhabited by people of the same class, the same income, the same age group ... a low-grade uniform environment from which escape is impossible" (cited in Ferrer and Navarra 1993, 15). Postwar suburbs such as Levittown were famously satirized in the lyrics of the Malvina Reynolds song "Little Boxes," in which the houses were said to be "all made out of ticky-tack" and "all look just the same." Yet Levittown is a far cry from the suburban slum predicted by Mumford, and today one would be hard-pressed to find any two houses that look the same.

What is interesting about Levittown is that the houses were designed with physical transformation in mind. Levittown houses were built with unfinished attics. Levitt actively encouraged homeowners to finish their attics and arranged for classes in the community centers to teach the skills necessary to do so. Peter Bacon Hales notes that the "barn-raising quality of the language surrounding these home improvements was palpable and calculated. It was also prophetic. For Levittowners banded together to work on each other's cars or share babysitting responsibilities, and they banded together to build out their homes" (Bacon Hales n.d.). Levitt's genius was in designing for unfolding relationships in a manner that used constraint reflexively. The assembly line production of the houses lent itself to creative remodeling. Structural changes to the original designs reflected changes in the growth of families. Houses were remodeled to suit the needs of families at progressive stages in their lives (Kelly 1993, 31). Houses grew with the families who owned them. Levitt's original designs were adapted for purpose in a specific historical and social milieu. Not only were the buildings "responsive," but their design also facilitated emergent social phenomena. Neighborhoods were invested with a conception of community, of commitment and belonging, which became desirable. Homeowners responded to the normative constraints of the home designs by actively reconfiguring their environment to suit their own needs (Kelly 1988, 46). The transformation of Levittown in turn worked to transform the residents, investing the development with a social capital that had, by the tenth anniversary of its opening, raised the value of both homes and residents. Levittown became a desirable location for middle-class Americans (52). Here we see an ideal realized through repurposing to suit family forms not necessarily anticipated by the designers, and social transformation brought about through the modification and personalization of the designed environment (Hadjiyanni and Helle 2008).

**Innovation and Improvisation**

We offer Levittown as an example of the kinds of processes through which users engage with cultural artifacts. Long ago, Claude Lévi-Strauss distinguished between (Western) engineers who make designs for particular purposes, follow plans, and produce specific components to realize their aims, and non-Western *bricoleurs* who adapt and modify existing elements (be they material or conceptual) to cope with new and changing situations (Lévi-Strauss 1966, 11). The human capacity for such adaptation has been understood as also present in modern societies, and as Sennet writes, "Many technology transfers that were meant to be merely rote applications of one procedure to another became illuminating just in this stage; there was something fuller and more manifold in the initial procedure than had been assumed" (2008, 211).

Taking an analogical leap from the adaptive modification of organic species to technical change, Tim Ingold has made a case for the human capacity for reuse as a logical extension of evolutionary change. Turning away from Lévi-Strauss's conception of bricolage, Ingold proffers the notion of "exaptation," by way of which "structures that may have evolved for one purpose are co-opted for quite different functions for which they happen to come in handy," as, for example, "the mammalian ear is derived from a part of the jaw of the fish" (Ingold 1997, 119). Here and in later writings, (e.g., Hallam and Ingold 2007), Ingold links this to the capacity for improvisation: "Most if not all of what we call invention, in the technical sphere, seems to involve a process of exaptation—of hitting on new uses for old things—[which] then condition subsequent processes of refinement."

One might say, following Lévi-Strauss's terms, that Levittowners were *bricoleurs* par excellence. Certainly there was a recombination of elements, materials, and ideas. Yet we would argue, following Ingold, that this was more than the reconstitution of a bounded set of elements, be they material or conceptual (cf. Derrida 1978). Levittown, uniformly diverse in a society that values individual autonomy and self-reliance, is an example of a common capacity for improvisation that is a consequence of social forces and values, of interaction with the material world, and with the agency of the designer. This repurposing is a generative moment that has the potential to go beyond the sum total of its parts. It is not reducible to the individual but expressive of historically contingent relations with things as well as people. Power relations reflected in the design of Levittown, including assumptions about race, gender, family, and the way that lives should be led, were made durable in the designed environment (Latour 1991). To the extent that the expansion and remodeling of homes engaged with and subverted Levitt's conception of the ideal family, we would argue that innovation was a version of reuse that transcended the original parts but did not involve the conscious, preplanned engineering of specific elements.

Innovation, especially in business management literature, is often defined as the successful implementation of alternative or creative thinking and of managed diversity in organizations. Creativity in these models appears in a linear relationship to innovation (Dionne 2004; Godin 2006). Hallam and Ingold, however, differentiate between innovation and improvisation. Often, they argue, the perceived difference between the two hinges on the way in which creativity is characterized. Improvisation characterizes creativity by way of its processes, whereas innovation characterizes creativity by way of its products. Thus to

> read creativity as innovation is ... to read it backwards, in terms of its results, instead of forwards, in terms of the movements that give rise to them. This backwards reading, symptomatic of modernity, finds in creativity a power not so much of adjustment and response of a world-in-formation as of liberation from the constraints of the world that is already made. (2007, 3)

Attempting to move beyond this linear model, they challenge the notion that creative improvisation is exercised against the conventions of culture and society. Rather, improvisation and creativity "are intrinsic to the very processes of social and cultural life" (2007, 19). Following the line of this argument, one might conclude that the transformation of Levitt's original vision of the ideal family, reflected in the environment that he designed, was less an attempt to move beyond stricture and more an expression of specific social capacities in interaction with technical deployment (Strathern, this volume).

Put another way, creativity and improvisation are expressions of common capacities that do not necessarily move against structure. We agree in principle that the notion of innovation is limited in its ability to engage with the creative processes of production that are characteristic of social life and cultural form.[4] And that is because, unlike the contributions gathered here, attention to innovation is all too seldom a consideration of the inherent creativity of social life. An attention, exemplified in what follows, to the politics of design and the results of cross-cultural or cross-sector knowledge encounters promises a liberation from the notion of creativity and design as the province of expert specialists. It also takes seriously the vital contributions possible where interaction is allowed for, and sensitivity to the processes of emergence and transformation provide durable components that can be repurposed. We begin with a story, told in poetic form, of the power of locality in shaping what is appropriate.

**Notes**

1. We use the acronym ICT (information and communications technology) rather than the terms "digital" or "digital media" intentionally to blur the distinction between digital and other technologies of knowledge production and reproduction.

2. For detailed information about bride wealth on the Rai Coast, see Leach 2003.

3. Levitt's utopian suburban dream was a strictly segregated vision of an ideal white community that had no place for black Americans. The rental agreement and the later restrictive covenant prohibited properties from being occupied by anybody except members of the "Caucasian race," and the whites-only policy defined the racial demographics of Levittown long after the struggle for equality in the United States had legally been won (see Bacon Hales 2004).

4. We would note, however, that in attempting the conceptual recuperation of improvisation as a means of eliciting the processes by which creative human capacities shape and are shaped by the lived-in world, there is a danger of neglecting the importance of power relations in design. This is a deficit that the contributors to this volume speak to directly.

## References

Agre, P. E. 1997. *Computation and Human Experience*. Cambridge: Cambridge University Press.

Bacon Hales, Peter. 2004. *Levittown's Palimpsest: Colored Skin*. http://tigger.uic.edu/~pbhales/race.html.

Bacon Hales, Peter. n.d. Unfinished: Expanding and decorating Levittown. http://tigger.uic.edu/~pbhales/Levittown/Decorating.html (accessed October 18, 2010).

Barab, S. A., M. K. Thomas, T. Dodge, K. Squire, and N. Markeda. 2004. Critical design ethnography: Designing for change. *Anthropology and Education Quarterly* 35 (2): 254–268.

Bell, G., M. Blythe, and P. Sengers. 2005. Making by making strange: Defamiliarization and the design of domestic technologies. *CM Transactions on Computer-Human Interaction (TOCHI)* 12 (2): 149–173.

Benkler, Y. 2006. *The Wealth of Networks: How Social Production Transforms Markets and Freedom*. New Haven: Yale University Press.

Bijker, E. W., and J. Law. 1992. *Shaping Technology/Building Society: Studies in Sociotechnical Change*. Cambridge, MA: MIT Press.

Bloor, D. 1991. *Knowledge and Social Imagery*. 2nd ed. Chicago: University of Chicago Press.

Bolter, J. D., and R. Grusin. 1999. *Remediation: Understanding New Media*. Cambridge, MA: MIT Press.

Broers, A. 2005. *The Triumph of Technology*. Cambridge: Cambridge University Press.

Brown, M. 1998. *Practical Reason: On the Theory of Action*. Stanford: Stanford University Press.

Brown, M. 2003. *Who Owns Native Cultures?* Cambridge, MA: Harvard University Press.

Burrell, J. 2009. The fieldsite as a network: A strategy for locating ethnographic research. *Field Methods* 21 (2): 181–191.

Button, G. 2000. The ethnographic tradition and design. *Design Studies* 21 (4): 319–332.

Castells, M. 1996. *The Rise of the Network Society*. Oxford: Blackwell.

Crook, T. 2007. Figures seen twice: Riles and the modern knower. In *Ways of Knowing*, ed. M. Harris. Oxford: Berghahn.

Davis, R., and J. Leach. 2012. Recognising and translating knowledge: Navigating the political, epistemological, legal, and ontological. *Anthropological Forum: A Journal of Social Anthropology and Comparative Sociology* 22 (3): 209–223.

Derrida, J. 1978. Structure, sign, and play in the discourse of the human sciences. In *Writing and Difference*, trans. Alan Bass. London: Routledge & Kegan Paul.

Dionne, S. D. 2004. Social influence, creativity, and innovation: Boundaries, brackets, and non-linearity. In *Multi-level Issues in Creativity and Innovation*, ed. D. Mumford, S. T. Hunter, and K. E. Bedell-Avers, 63–74. JAI Press.

Ferrer, M. L., and T. Navarra. 1993. *Levittown: The First 50 Years*. Arcadia.

Gell, A. 1998. *Art and Agency*. Oxford: Oxford University Press.

Ghosh, R. A., ed. 2005. *CODE: Collaboration and Ownership in the Digital Economy*. Cambridge, MA: MIT Press.

Godin, B. 2006. The linear model of innovation: The historical construction of an analytical framework. *Science, Technology, and Human Values* 31 (6): 639–667.

Gurstein, M. 2000. *Community Informatics: Enabling Communities with Information and Communications Technologies*. Hershey, PA: Idea Group Publishing.

Hadjiyanni, T., and K. Helle. 2008. Re/claiming the past: Constructing Ojibwe identity in Minnesota homes. *Design Studies* 30:462–481.

Hallam, E., and T. Ingold. 2007. Creativity and cultural improvisation: An introduction. In *Creativity and Cultural Improvisation*, ed. E. Hallam and T. Ingold, 1–24. Oxford: Berg.

Harris, M., ed. 2007. *Ways of Knowing: New Approaches in the Anthropology of Experience and Learning*. Oxford: Berghahn.

Heckler, S., ed. 2009. *Landscape, Process, and Power: Re-evaluating Traditional Environmental Knowledge*. Oxford: Berghahn.

Hess, D. J. 2001. Ethnography and the development of science and technology studies. In *Sage Handbook of Ethnography*, ed. P. Atkinson, A. Coffey, S. Delamont, J. Lofland, and L. Lofland, 234–245. Thousand Oaks, CA: Sage.

Horst, H. A., and D. Miller. 2006. *The Cell Phone: An Anthropology of Communication*. Oxford: Berg.

Ingold, T. 1997. Eight themes in the anthropology of technology. *Social Analysis* 41 (1): 106–138.

Kelly, B. M. 1988. Learning from Levittown. *Long Island Historical Journal* 1(1): 39–54.

Kelly, B. M. 1993. Little boxes, big ideas. *Design Quarterly* 158:26–31.

Latour, B. 1991. Technology is society made durable. In *A Sociology of Monsters? Essays on Power, Technology, and Domination*, ed. J. Law, 103–131. Oxford: Blackwell.

Latour, B. 1993. *We Have Never Been Modern*. Cambridge, MA: Harvard University Press.

Leach, E. 1968. *A Runaway World? The 1967 Reith Lectures*. London: British Broadcasting Corporation.

Leach, J. 2003. *Creative Land: Place and Procreation on the Rai Coast of Papua New Guinea*. Oxford: Berghahn Books.

Leach, J. 2005. Being in between: Art-science collaborations and a technological culture. *Social Analysis* 49 (1): 141–160.

Leach, J. 2010. Intervening with the social? Ethnographic practice and Tarde's image of relations between subjects. In *The Social after Gabriel Tarde*, ed. M. Candea. London: Routledge.

Leach, J. 2012. Leaving the magic out: Knowledge and effect in different places. *Anthropological Forum: A Journal of Social Anthropology and Comparative Sociology* 22 (3): 251–270.

Leach, J. 2013. Choreographic objects: Contemporary dance, digital creations, and prototyping social visibility. In *Prototyping Cultures: Art, Science, and Politics in Beta*, ed. Alberto Corsin-Jimenez. Special issue, *Journal of Cultural Economy*.

Lévi-Strauss, C. 1966. *The Savage Mind*. Oxford: Oxford University Press.

Lievrouw, L. A. 2006. New media design and development: Diffusions of innovations vs. the social shaping of technology. In *Handbook of New Media*, ed. L. A. Lievrouw and S. Livingstone, 246–265. London: Routledge.

Lievrouw, L. A. 2011. *Alternative and Activist New Media*. Malden, MA: Polity.

Marchand, T., ed. 2010. *Making Knowledge: Explorations of the Indissoluble Relation between Minds, Bodies, and Environment*. Special Issue, *Journal of the Royal Anthropological Institute* 16 (S1): S1–S21.

Millen, David R. 2000. Rapid ethnography: Time deepening strategies for HCI field research. In *Proceedings of the Third Conference on Designing Interactive Systems: Processes, Practices, Methods, and Techniques*, 280–286. New York: ACM Press.

Morphy, H. 1991. *Ancestral Connections: Art and an Aboriginal System of Knowledge*. Chicago: University of Chicago Press.

Morphy, H. 2007. *Becoming Art: Exploring Cross-Cultural Categories*. London: Berg.

Morphy, H., P. Deveson, and K. Hayne. 2005. *The Art of Narritjin Maymuru*. Canberra: ANU E Press.

Ong, A., and S. J. Collier. 2004. *Global Assemblages: Technology, Politics, and Ethics as Anthropological Problems*. Oxford: Blackwell.

Rubio, Fernando Dominigez, and Patrick Beart, eds. 2012. *The Politics of Knowledge*. London: Routledge.

Runciman, W. G. 1983. *The Methodology of Social Theory*, vol. 1: *A Treatise on Social Theory*. Cambridge: Cambridge University Press.

Sennet, R. 2008. *The Craftsman*. London: Penguin.

Shapin, S., and S. Schaffer. 1989. *Leviathan and the Air Pump: Hobbes, Boyle, and the Experimental Life, Including a Translation of Thomas Hobbes, Dialogus Physicus de Natura Aeris, by Simon Schaffer*. Princeton, NJ: Princeton University Press.

Sillitoe, P. 2009. *Local Science vs. Global Science: Approaches to Indigenous Knowledge in International Development*. Oxford: Berghahn.

Sillitoe, P., A. Bicker, and J. Pottier. 2002. *Participating in Development: Approaches to Indigenous Knowledge*. ASA Monographs 39. London: Routledge.

Strathern, M. 2004. Laudable aims and problematic consequences, or The "flow" of knowledge is not neutral. In *Economy and Society* 33(4): 550–561.

Strathern, M. 2010. What's in an Argument? Preliminary Reflections on Knowledge Exchanges. UBC Vancouver Harry Hawthorn Lecture.

Takeshita, Chikako. 2001. Bioprospecting and its discontents: Indigenous resistances as legitimate politics. *Alternatives: Global, Local, Political* 26 (3):259–282.

Tonkinwise, C. 2005. Is design finished? Dematerialisation and changing things. In *Design Philosophy Papers, Collection Two*, ed. A. M. Willis, 20–30. Ravensbourne, Australia: D/E/S Publications.

Wagner, R., and A. D. Wagner. 2010. *Images of America: Levittown*. Charleston, SC: Arcadia.

Weber, S. 2004. *The Success of Open Source*. Cambridge, MA: Harvard University Press.

Weiner, J. F. 1995. *The Lost Drum: The Myth of Sexuality in Papua New Guinea and Beyond*. Madison: University of Wisconsin Press.

Weiner, J. F. 2001. *Tree Leaf Talk: A Heideggerian Anthropology*. Oxford: Berg.

Wilson, L., and D. Leitner. 2009. *Naked Ambition*. Critical Ethnographic Evaluation of Arts Council IT for the Arts Initiative. Cambridge University Technical Services. http://www.getambition.com/resources/final-report-naked-ambition.

## 2  Liminal Futures: Poem for Islands at the Edge

Laura Watts

### Introduction

The following prose poem was a response to my first month of ethnography in Orkney, an archipelago off the northeast coast of Scotland. I wanted to gather my experiences and evoke them in a way that academic prose could not do. At the time, I was living and working with people who imagine and design future technologies in the islands. My interest was, and is, in how location and landscape affect the way the future is imagined and made. How are futures made differently in different places? Why are certain landscapes and places regarded as centers of innovation, and others as peripheral, and how might relocating our attention help refigure "innovation" and future-making? The future is not out there, floating on some temporal horizon, but, following approaches in social studies of science and technology, must be made in strategy meetings and standards documents, in practices of design and demonstration (Bowker and Star 2000; Suchman 2007). The future is always local to its sites of ongoing rehearsal and production; as with other forms of knowledge, futures are situated (Haraway 1991).

Orkney is a low-lying, relatively fertile archipelago of around twenty thousand people living among seventy scattered green and heather islands. Off the northern coast of Scotland, they are closer to the Arctic Circle than to London. These people and places are too often considered remote, at the northern periphery of the United Kingdom and Europe, and far from the usual centers of high-tech industry and future-making; beef farming is Orkney's largest industry. Why are these islands, then, also an international center for marine renewable energy research, and before that for wind energy research? Why have the environmental resources of Orkney, its powerful tides, waves, and wind, led to such world-renowned activity? Why do people from the global renewable energy industry, from Taiwan to Silicon Valley, visit this archipelago in droves (Watts, 2012)?

I came to these islands as an effect of prior research in the Thames Valley near London, a landscape of high-tech business parks, and a classic center of UK innovation. There I spent time with those making the future for the mobile telecom industry. Their future was often imagined as ubiquitous and pervasive, a future of unimpeded global access to information. But the landscape in which this future was located was one of dense mobile phone networks, cities lit by buried fiber optics, flat enough for radio signals to propagate with relative ease, where smooth motorway tarmac and Heathrow flight corridors moved bodies and bytes at speed, and relative wealth created an experience of pervasive access to information for many (Watts 2008). The local landscape for this global future of "anyone anywhere anytime" connectivity was one of already pervasive information access. The mobile telecom industry future of worldwide "always on" information seemed to be more a replication of their everyday local experience; a copy-and-paste of their sociotechnical landscape.

Some I spoke with in the mobile telecom industry were frustrated by this limited imaginary, which seemed to create substantial material and economic problems, such as vast debts from radio spectrum auctions (Cheng, Tayu, and Yu 2003), and a myopic fixation on ever-increasing bandwidth (sometimes characterized as Moore's Law; Watts 2008). They questioned how the industry might reimagine its future. How to do it otherwise?

So I wondered, how might the future be reimagined and redesigned if the landscape were different? If knowing is an effect of moving through a particular landscape (Ingold 1993; Turnbull 2002; Chambers and Turnbull, this volume), then so too are accounts, visions, and versions of the future. For example, what futures might be made moving through a landscape whose topography of hills and sea resists the propagation of mobile phone signals and optical fiber, a landscape whose temporality is not of newly erected mobile phone masts but of enduring five-thousand-year-old Neolithic stone monuments? How might ICT futures be imagined and made differently in such a different place?

Orkney is such a landscape. It has its own particular high-tech industries, its own particular approaches to making futures and future technology. Perhaps it is these distinctive futures, imagined and designed through these distinctive islands, that draw so many global visitors each year (tourists, politicians, venture capitalists, engineers, ethnographers). It was these *situated futures* I sought to understand during my ethnography in the islands. How were these futures made at the geographic edge, liminal futures that invited both local and international participation, that inspired so many, and perhaps resituated the so-called leading edge of the future?

# Liminal Futures

## Part I: Insistent Infrastructure

Hailstones
roaring down and across
the slab rectangles of concrete runway;
my twin-propeller plane is towed for repair.
Flying to Orkney is never certain.
I have heard tales of ghostly passengers
haunting the departure hall of Aberdeen for days,
waiting for winter weather to pass.

I land on Orkney in torrential sleet,
mountain summit wind
blazing,
whipping my car door near off its hinges,
then, pressurizing me in.

The supermarket shelves are laid waste.
Locusts ahead of the storm have cleared two aisles.
Not even a carrot.
For if the boat does not sail,
then the shelves are not filled.
Food and fuel,
rolling on and rolling off the ferry,
are barometers;
piled "high" only by the will'o the sea,
that most treacherous of waters,
ridden with whirlpools:
the Pentland Firth collision
between Atlantic and North Sea tides.

The worldly rise and fall of food and fuel are visible here.
Petrol and carrots register their weighty resistance to transport
in the balance of empty shelves.

"You are connected to the weather here,
connected to the supply chain," says a local librarian.
"There is a sense of the interconnectedness of things."
"Western consumerism is tempered," she says.
There is clear green water always in sight,
between you and your wants;

a ferry or two,
between want and can have.

"We're at the end of the supply chain,"
says a start-up director,
renewable energy maker;
his company reversing the bearing of supply and demand:
north to down south,
not south to up north.

Food and fuel,
weather and wind,
people and property:
their hard-world, hard-won relations are
labeled and marked clearly
in the high cost of electricity,
in every boat-carried kettle, pencil, bright summer dress;
in the slow, barely megabyte, broadband connection,
(but "good view," says the marketing writer,
a remote-presence working from home;
"sod the broadband," he jokes).

The infrastructure of Western world living
is laid bare and skin-close:
an insistent touch of telecoms and transport;
the smell of energy on the wind.
(No urban sprawl to hide the passage of cable and coax,
no dense population to excuse the economic accounts,
no easy flat city to roll out the fiber.)

The National Grid cable,
an electrical lifeline
under the sea,
over the hills and protected peat moorland,
is the murmur of islanders,
the talk on the ferries.
A new cable is needed, capacity reached.
But where should it flow?
And many people speak,
have a future they see:
Tim worries for the archaeology;
Keith for the view;
Martin for peak oil,

and Annie for folk,
those who live, work, and die here.

Grant points to the problem of
infrastructure centralization;
the UK's postwar electrical system for urbanization.
But archipelago Orkney is *de*central,
distributed,
urban distant;
and so has one of the country's first
local
self-determined
power management systems.

And it seems as if the future,
the one where people care for their networks,
for their infrastructure, supply and demand,
that future has come early to Orkney.

## Part II: Self-Sufficiency

So this is not life in the past,
not a honeydewed heritage,
"not Orkney in a jar of formaldehyde,"
says the archaeological curator.
"Not fossilized," says the start-up director
of a potential wind turbine farm.
"We must be ancient and modern," he continues to call.
"It's a living landscape.
We make a living here ..."

... from farming and producing:
biogas, electricity,
beef, cheddar, and fudge.
Producing and farming
is a futures way of living
(as Jo Vergunst reminds me):
a care for the next generation of cattle, community, and crop.
Some here cite a millennium of farming descent;
generations of futurists,
farming for the future.
Since the Vikings, go the tales,
part myth, part gene.
"You belong to the land,

and the land belongs to you," says a local archaeologist.
Hefted
is a word I hear;
a people made land,
a land made people:
making the future.

"I ask you to imagine
how you see the world in the future,"
calls the start-up director,
farmer of wind, earth, and time.
He speaks,
compelling the crowd and community:
"I believe hydrocarbons will become rare in everyday use …
We're not the first place the government will send energy
in energy-scarce times."
He speaks of a storm-force future
when the energy boat does not run.

"It is the sea which has contributed to self-sufficiency,"
says the islands council
in its promotional brochure.
These feisty waters hold depths of invention,
a storehouse of marine power,
fish stock, and tales.
So tethered in Orkney are world firsts in
tide and wave power;
testing marine energy feasibility
in the wild
waters.

And Scotland's first locally owned wind turbine
stands proud on its isle;
joined now by others,
more planned,
futures too numerous to count,
funded, owned, and contested
by folk hereabout.

Orkney is "Initiative at the Edge"
(as a government fund names its far reaches).
This is life at the edge
of Western world living,
sharp,

in focus,
cutting-edge;
cutting its own way.

And it seems as if the future,
the one where people care for their future,
long local, practical, self-sufficient,
that future has come early to Orkney.

## Part III: Modest Innovation

But here is a tension,
in the viscous flow of oil:
energy poverty
is greater here than elsewhere in the country.
And yet on one island,
burning orange-strong in the wind,
is the Flotta oil terminal flare.
Ten percent of the UK's energy
lands from North Sea oil fields
and is taken from there.
The smell of energy is on the wind,
in the carbon of the earth,
in the uranium of the stone.
"Orkney is in every sense an energy island,"
says the renewable energy forum.
(It's a land-made-people,
people-made-land,
made future.)

"We aim as an island to be 100 percent renewable
by 2012,"
says the director of a biogas firm;
and son-in-law of the farmer
of a trial site for the fuel.
"You know all the board members,"
says the director of a wind turbine firm,
and father-in-law and husband
to others on the company board.
You see, it's all in the family,
communal and relational.
It's trust in the network,
the community bond,
(as Michael Lange says).

A high-tech company director here names it:
transparency,
integrity.
You are beholden to every word spoken,
for there are no casual encounters,
no conversations with strangers,
no hiding in a city of blank faces, blank words.
What is said is remembered,
repeated,
returned.
So it's quick to get a decision here,
I'm told with a grin,
for everyone knows everyone
(for good and for gossip).

But what's remembered is holding the door open,
as much as a global expertise.
"We include lepers but exclude arrogance,"
says the local ecologist.
For there is no need to shout.
Quiet,
resourceful,
these are the terms
the local high-tech director repeats
(in comparison with American West and East coasts).
"Right, let's get on with it,"
is the response to a need for new futures,
says the island council solicitor.
So farm and food waste becomes biogas, biofuel;
the smell of chip fat on the car-carried wind.
It's a land-made-people,
people-made-land,
getting on,
making the future.

Modest innovation,
that's what I sense.
Not entrepreneurship,
a term that's disliked
as too "bigsy,"
too big for your boots,
too self- not community centered;
(it's the next generation of cattle, community, and crop).

# Liminal Futures

And it seems as if the future,
the one where people care to join hands,
get on and make something new happen,
that future has come early to Orkney.

## Part IV: Mutable Futures

Hailstones
fall on wet mush and March snow.
I stand in the center,
the fulcrum of Orkney:
the Neolithic stone circles of
Brodgar, Stenness.
Sandstone mica glows soft
with spring shadows.
The grass, stone, and sea
are a circular world in
green, gray, and brown;
the color of tundra
and fast-tracking storms.

On this horizon of five thousand years
of building and dwelling
is a wind turbine farm,
imagined,
unnegotiated,
unformed,
heartfelt.
This landscape is living,
unfossilized,
unfixed.

In a community hall the crowd gathers
to debate
three other wind turbines.
Will they turn, over the sun?
Some speak for the peat bog,
high on the hill.
Some speak for the ravens.
One for the National Grid.
One for the skyline,
the tourists and their trade.
Another for the shadows, strobing her house.
And all speak of the future,

so many conflicting.
For the future is open,
mutating,
remaking.
The turbines shifting,
dissolving
in quiet talks over farmhouse
coffee and cake.
These turbines are open,
not just in material shape and design,
but in ownership and presence;
a whole part of a land-made-people,
a people-made-land.

And it seems as if the future,
the one where people remake their own future,
make it mutable, fluid,
that future has come early to Orkney.

**Part V: Early Adapters**

Grant demos the tools of his mutable trade:
a Virtual Terrain Program
that models wind energy;
makes visible, malleable, commensurable,
the many voices and species that speak.
In viewsheds,
acoustic impact zones,
and statutory sites:
archaeological monuments speak,
birds speak,
the Orkney vole and otters speak.
Data spreads speak for the hen harriers
(from Jules watching and counting their flight),
for the grass type
beneath my virtual feet.
There are data for the wind strength,
for the heather and the houses;
for the laws of the landscape,
the economics of business.

It's all here at the touch of a slider,
a dial to turn up or turn down;
tuning in to a future,

working to resonate with the possible.
These are songs of the future,
scored on CD-ROM;
voices composed into visuals,
not for comparison but for imagination.
Songs of the future,
scored into a virtual world,
data voices open to creative interpretation;
a world designed to be altered,
made otherwise.

This is high-tech industry at the edge
of Western world living;
cutting-edge,
cutting its own way;
an early adapter,
not early adopter.

And it seems as if the future,
the one where technology is designed to be
open, personal, adaptable,
that future has come early to Orkney.

## Reprise

Orkney is life at the edge
of Western world living;
sharp,
hard,
demanding.
"We are a place where technologies are trialed."
"A technological test bed,"
says a high-tech company director;
testing remote-working,
renewable energies,
renewable pasts.

And it seems as if the future,
the one where innovation *is* at the edge:
difficult, distributed, decentralized,
that future has come early to Orkney.

April 2008

## Discussion

*The future has come early to Orkney.* This future is shorthand for an imagined future that is designed, developed, and made in the present, as outlined in the introduction. It is a design practice with particular qualities in Orkney that would seem to be of interest to ICT developers elsewhere. This discussion will not translate or speak for the prose poem; it does its own work. However, I do want to explore the refiguring of this future in the poem: the distinctive way I encountered the future being imagined and developed in Orkney. And I want to consider the implications of this refigured future, and its making, for more normative accounts of ICT design and development—for example, implications for future-making in the mobile telecom community around London, highlighted in the introduction.

*The future has come early to Orkney.* How to make sense of such a claim, one that is an explicit interruption of the usual story: well-known urban centers of innovation are somehow technologically ahead, on the leading edge of the future (e.g., London, Silicon Valley, Tokyo), while rural peripheries languish behind such innovation centers. The poem's claim implicitly evokes and plays with well-known critiques of technological determinism and linear technological evolution (Latour 1987; Bijker and Law 1992; Latour 1993). But its mirrored reflection of remote islands, rather than urban centers, as technologically ahead has some precedent. Fragile and remote islands whose populations are acutely aware of their own potential transience in time, whose finely tuned ecosystems are hypersensitive to change—such places have long been named as "advance indicators or extreme reproductions of what is future elsewhere" (Baldacchino 2007a, 9), and as "the first, the harbingers, the pioneers, the miner's canary" (Baldacchino 2007b, 166). As former UN Secretary-General Kofi Annan put it in his discussion of Small Island Developing States, "Islands are frontline zones where many of the main problems of environment and development are unfolding" (Baldacchino 2007b, 7). Orkney is a pioneer, a living reproduction of what is only an imagined future in other locations as an effect of its fragile archipelago landscape, and by *landscape* I mean both natural and cultural, both the people and the place. The fragility of Orkney is an everyday experience for the people who live there. The dependent infrastructures of contemporary living, forgotten and literally buried in urban places, are visible and embodied in the weather-reliant ferries, in the occasional electricity blackouts, in the "not spots" of absent broadband and mobile phone signal; *insistent infrastructure*, as the poem names it. As an effect of such tenuous connections, depopulation is a pressing concern for the twenty or so inhabited islands, some with fewer than a hundred people. A sustainable future is not something abstract talked about only by national politicians or global business, not a distant worry that can be packaged and shelved as a vague matter of "global climate change." Here it is a matter of community survival, next year and every year. The future, especially future infrastructure technology, is an everyday concern, talked about over farmhouse coffee and cake, as the poem suggests; everyone is a participant. In contrast, the government in London talks about such Orkney experiences as if they lie in the future: "A high-carbon world is one with more extreme weather, where we and our children are faced with the costs of adapting the way we live and the infrastructure and systems that support us. We must face up to these challenges and make the necessary investment to move to a low-carbon economy now" (DECC 2009, 18). Orkney is a frontline zone where this future of local adaptation to extreme weather and limited infrastructure, local investment in up-and-running renewable energy projects, is already lived. It is more than a matter of landscape agency or environmental determinism. The specificity of how that low-carbon future has been made and lived, the qualities of design and development in Orkney that perform such local adaptation (named in the poem as *self-sufficiency*, *modest innovation*, *mutable futures*, and *early adapters*), is therefore of wide relevance.

Reconfiguration could perhaps be one way to characterize Orkney approaches to future-making, the reconfiguration of not just technical parts but also social and environmental parts to "get on and make something new happen." The renewable energy futures in the poem are created through ongoing social, technical, and environmental negotiation and contestation; the heterogeneity needed to make future technology is obvious here (Law 1992). Wind turbines are not imported and simply installed but understood as reconfiguration projects that, with great care and personal investment, must be translated and negotiated through intense discussion and community debate. Some wind turbine projects are sufficiently mutable and can establish ongoing relations with the community, the archaeology, the wildlife, the peat, and so on, while others cannot and so do not happen. The biogas start-up company and its biofuel technology are embedded in, and integral to, familial relations, cattle that produce slurry for fuel, and an island community and its strategy to become energy self-sufficient. The Neolithic archaeology of stone circles and the future archaeology of wind turbines are seen as concurrent parts of the landscape, which have to be reconfigured and woven together in the hard work of making sociotechnical and environmental relations, if they can be woven together at all (there is no guarantee of relationality). That the future is situated, that technology must be situated socioculturally and environmentally for it to be sustainable, in the double sense of both enduring over time and as enrolled in a strategy for self-sufficiency, is unquestioned. The work of situating technology, that is, trying to get technology designed elsewhere to operate and keep operating in Orkney, is just too hard to be ignored. Getting ICTs designed for different places to operate here means *hacking* the relational connections between technology and community, as much as with the stormy weather and disparate archipelago landscape. Orkney resists technologies designed for cities and suburban landscapes, making visible the reconfiguration work that always happens as technologies are supposedly "rolled out" to new places. This reconfiguration work has been well discussed elsewhere as a necessary aspect of technology design, development, and use (e.g., Suchman 2002). Ethnographic studies have shown how local improvisation by "users" plays an integral part in making a technology work as it moves from place to place (see de Laet and Mol 2000), and how these so-called users can be more helpfully understood as participants in the design process, or as parts of the sociotechnical infrastructure in which a technology, such as a wind turbine, is embedded (Star 1999, 2002). Orkney's resistance to rolled-out energy or telecom networks, its resistance to copy-and-paste infrastructures imagined and designed elsewhere, is not a resistance to innovation (as is perhaps too often thought) but precisely its opposite. It is a resistance that both *generates* innovation, in the sense of generating intensive local improvisation to make things work, and *invokes* innovation, in the sense of invoking designers to collaborate with the islands and so make their designs and futures work "anywhere." Orkney's resistance is a call for design innovation, for designers to reimagine their technology embedded in a more diverse landscape, a landscape that is not behind but on the future front line.

However, reconfiguration as an approach to making the future is not heroic or romantic, not an inherent rural good over some urban bad. Archipelago self-determinism does not negate the important question of who determines. Although the Orkney vole and otter speak, although the farmer and archaeologist speak, although the different island communities speak, they do not all have an equal say. Relations, both familial and heterogeneous, are always ongoing and always come with obligations (see Strathern, this volume). Negotiation and contestation are never rational or equitable. The future in Orkney is an effect of a morass of silent obligation and "wrong" decisions for some, despite intentions otherwise. But it is that morass of obligation, the dense knit of social, technical, and environmental relations and exchanges—people committed to, and part of, the place, as much as part of extended Orcadian families and island communities (Forsythe 1982; Lange 2007; Lee 2007)—it is these relations that produce a capability to act fast, to have a high-speed response: "It's quick to get a decision here, for everyone knows everyone." So all it can take is a few brief telephone calls to bring together the island council representative, the company director, the farmer, the designer, the academic, to put an idea into practice that afternoon, because the relations are already woven together; a commitment and obligation to the islands' future is already there. This contrasts with the popular conception of rural life as slow paced, versus the speeding temporality of urban living. Orkney is not slow to act but the converse, able to swiftly reconfigure and transform relations between people, places, and things, and so enact futures imagined elsewhere. This capability to make rapid rearrangements of relations, to reuse, is recognizable as Orkney's *self-sufficiency*. This self-sufficiency is the reconfiguration of existing relations: chip fat is also biofuel. But it is not resource management, for things always push back, if not outright talk back; not every relation goes. Self-sufficiency involves a certain relational creativity, putting things together in new ways: old chip fat from the island chip shop becomes new biofuel; the old familiar car becomes the new island-converted biodiesel car. In Orkney the "new" is made by reusing and reconfiguring the "old"; invention is a matter of rearrangement, not technical novelty (Barry 1999). This is not due to some romantic sentiment but rather a necessary effect of its remote islandness and the insistence of the infrastructure; flows of new goods, people, data, and energy are not unending but stop in bad weather, so you have to make your own new things. For this reason, remote small islands often evoke sustainable futures (Kerr 2005). Future-making in Orkney is self-sufficient, sustainable, creative, and fast paced as an effect of its particular location, its particular geography and history, its particular relations and obligations between people, places, and things.

Orkney future-making, in the form of reconfiguration, also resists heroic inventors and competitive entrepreneurialism ("too bigsy") often associated with high-tech industry. Instead it promotes communal endeavor, shared participation, and self-effacement. This modesty is not due to some naturalized goodness in island or rural communities but is an effect of island obligations to share, for benefits to be communal; what matters is "a care for the next generation of cattle, community, and crop." Modest innovation is an effect of island living and its frontline concern with the future survival of the community, not the individual. Such modesty creates a future with a politics different from more normative ICT industry futures, where celebrations of the lone, often male, inventor or designer are common (Wajcman

1991). In addition, such modesty has a temporality different from more heroic versions of future-making. There is a care in Orkney for the longer term, for the next generation of the community and place, and a sense of long-term continuity over millennia ("five thousand years of building and dwelling"). In contrast, designers in the mobile telecom industry, as in other high-tech industries, often have a shorter-term care for the next generation of products or the next economic quarter as part of their own obligations to their corporate community.

I am not arguing for a copy-and-paste of Orkney future-making practices to urban centers of design and innovation. Neither am I arguing that Orkney approaches to future-making are without similarity elsewhere; creative relationality, for example, is by no means limited to these Scottish islands (Ingold and Hallam 2007). But, as I hope this discussion has made clear, Orkney reconfiguration approaches to future-making are situated in Orkney. Its practices are integral to its location, to the islands, the people, and the place. Just as copy-and-paste approaches to designing new technology in London do not work when "rolled out" in Orkney, the opposite must also be the case. Perhaps the crucial difference is that those who participate in making the future in Orkney appear to understand its locatedness: their everyday experience involves reconfiguring technology to work in different islands in the archipelago; a wind turbine project in one island is not the same as in another (even if the technology has the same specification).

Given this refusal to just adopt or universalize Orkney reconfiguration approaches to design and development, what can be gained by future makers elsewhere? How might this discussion help mobile telecom industry designers reimagine their future otherwise?

First, the refusal to universalize is itself an important mechanism to make visible universal futures that are otherwise taken for granted, such as visions of ubiquitous computing (Dourish and Bell 2007). These futures are located in the social and technical landscapes where they are made, as much as Orkney futures. High-tech industry communities are just as situated, just as embedded in their own places and obligations, as those in island peripheries. Ubiquitous telecommunications, which require ever-increasing bandwidth and ever-increasing infrastructure rollout, are a particular future, imagined as an effect of a particular place. As an imagined future that leads to people making design decisions, it is no more or less global than the self-sufficient and reconfigured future made in Orkney; although they have substantive differences in power and influence, futures are not equal. To acknowledge that the future is situated in people and places is to acknowledge that there are many possible futures, in many places. Rather than one global solution, there are only ever local variants, ever more local versions that you cannot fully describe or document, no matter how many field sites you visit. Situated futures go all the way down. So rather than imagining and designing one universal future technology for all, how might a design be envisioned for people to redesign? How to make a future technology that is malleable and adaptable, designed for incorporation into other futures, designed for development by other places, places a designer or product manager never imagined? How to refigure ownership of the future so that this is possible, for example, open standards, open source, copyleft? What futures can be imagined by high-tech industry that are open to transformation by others, open to becoming embedded in other landscapes, rather than demanding subversion tactics to get them integrated and working?

Second, colloquial assumptions concerning peripheries and rural landscapes as behind the future are not necessarily the case. Instead there are some, perhaps unexpected, places (not necessarily outside one's national border; see also Nafus, this volume) where futures are made in different ways. This is not a simple matter of geography: not all islands are frontline zones, and not all urban business parks are sites of cutting-edge innovation. Attention to local practice, to how a future is made in a particular place, is necessary. The benefits of this attention are not the acquisition of new futures that can be applied to the latest design, however. Instead such attention can be the beginning of a collaboration between designers, both local and visiting, leading to a shared understanding: in practical terms, what works in that place; and in strategic terms, a more nuanced articulation of what counts as the future, innovation, and creativity (e.g., Leach 2004). The prose poem, its draft description of Orkney future-making, is itself an example of such a collaboration, one to which I am unashamedly committed.

Finally, the archipelago's ability to reconfigure what is at hand, to try things out, is also a quality that supports testing and trialing of new technology. The poem calls the islands an *early adapter*. Orkney has a practical expertise in getting things to work in its diverse and difficult landscape; it is a fast rewirer of relations. Moreover, in a landscape on the sustainable-future front line, designers and future makers also have an opportunity to test prototypes in a place that is already living the future, and where future technologies are an everyday, vivid concern. This opportunity, Orkney as an industry test site for futures, is already ongoing in the form of the European Marine Energy Centre (EMEC). Designers, politicians, engineers, mariners, and many others are collaborating with the islands, drawing on local professional expertise as well as the local environment, to imagine and reconfigure energy infrastructure and energy futures—from wave and tidal turbine devices to new environmental standards and electricity grid networks.

Orkney as a test site is not a version of the nefarious island laboratory, or island as *terra nullius* to be colonized and experimented on (Rainbird 1999). Rather, it is laboratory as collaboration, laboratory as extended community of people, places, and things, where a shared commitment is created to the futures under construction. It requires those who visit with their prototypes to understand that in testing them, in getting them working and learning from the islands, they become participants in making an Orkney future as well as their own; collaboration comes with obligations that go both ways. This is always the case when organizations work together, and there are well-established modes for doing so (e.g., non-disclosure agreements). However, partnership between diverse groups, such as between ICT design centers and island peripheries, requires a duty of care on both sides, and attention to the mutual benefits—attention to both futures.

*The future has come early to Orkney* is an opportunity for both the islands and high-tech industry to collaborate and to imagine and make their futures together. Orkney is an archipelago at the edge, both fragile and futuristic. The futures that might be designed there are liminal: they are at the limit of the possible, at the storm-force, spread-too-thin limit of infrastructure; and they are betwixt and between, formed and reformed from fluid and shifting relations, always being made anew. In Orkney the future can always be otherwise.

### Acknowledgments

Warm thanks to all in Orkney who shared their time and thoughts with such generosity, and in particular to everyone at Aquatera Ltd. who made me feel so at home. This research was part of The Leverhulme Trust project "Relocating Innovation: Places and Material-Practices of Future-Making" at the Centre for Science Studies, Lancaster University. Special thanks to Lucy Suchman and Endre Dányi, my colleagues on the project, who always supported, questioned, and inspired.

### References

Baldacchino, Godfrey. 2007a. Introducing a world of islands. In *A World of Islands: An Island Studies Reader*, ed. Godfrey Baldacchino. Charlottetown: Institute of Island Studies, University of Prince Edward Island.

Baldacchino, Godfrey. 2007b. Islands as novelty sites. *Geographical Review* 97 (2): 165–174.

Barry, Andrew. 1999. Invention and inertia. *Cambridge Anthropology* 21 (3): 62–70.

Bijker, Wiebe E., and John Law. 1992. *Shaping Technology/Building Society: Studies in Sociotechnical Change*. Cambridge, MA: MIT Press.

Bowker, Geoffrey, and Susan Leigh Star. 2000. *Sorting Things Out: Classification and Its Consequences*. Cambridge, MA: MIT Press.

Cheng, Joe Z., Joseph Z. Tayu, and Hsiao-Cheng D. Yu. 2003. Boom and gloom in the global telecommunications industry. *Technology in Society* 25:65–81.

DECC. 2009. The UK Low Carbon Transition Plan Executive Summary. Department of Environment and Climate Change. London. http://www.decc.gov.uk/publications.

de Laet, Marianne, and Annemarie Mol. 2000. The Zimbabwe bush pump: Mechanics of a fluid technology. *Social Studies of Science* 30:225–263.

Dourish, Paul, and Genevieve Bell. 2007. Yesterday's tomorrows: Notes on ubiquitous computing's dominant vision. *Personal and Ubiquitous Computing* 11 (2): 133–143.

Forsythe, Diana. 1982. *Urban–Rural Migration, Change, and Conflict in an Orkney Island Community*. London: Social Science Research Council.

Haraway, Donna. 1991. Situated knowledges: The science question in feminism and the privilege of partial perspective. In *Simians, Cyborgs, and Women: The Reinvention of Nature*, 183–201. London: Free Association Books.

Ingold, Tim. 1993. The temporality of the landscape. *World Archaeology* 25 (2): 152–174.

Ingold, Tim, and Elizabeth Hallam. 2007. *Creativity and Cultural Improvisation*. ASA Monograph. Oxford: Berg.

Kerr, Sandy. 2005. What is small island sustainable development about? *Ocean and Coastal Management* 48:503–524.

Lange, Michael A. 2007. *The Norwegian Scots: An Anthropological Interpretation of Viking-Scottish Identity in the Orkney Islands*. Queenston, Ontario: Edwin Mellen.

Latour, Bruno. 1987. *Science in Action: How to Follow Scientists and Engineers through Society*. Cambridge, MA: Harvard University Press.

Latour, Bruno. 1993. *We Have Never Been Modern*. Hemel Hempstead: Harvester Wheatsheaf.

Law, John. 1992. Notes on the theory of the actor-network: Ordering, strategy, and heterogeneity. *Systems Practice* 5:379–393.

Leach, James. 2004. *Creative Land: Place and Procreation on the Rai Coast of Papua New Guinea.* Oxford: Berghahn Books.

Lee, Jo. 2007. Experiencing landscape: Orkney hill land and farming. *Journal of Rural Studies* 23 (1): 88–100.

Rainbird, Paul. 1999. Islands out of time: Towards a critique of island archaeology. *Journal of Mediterranean Archaeology* 12 (2): 216–234.

Star, Susan Leigh. 1999. The ethnography of infrastructures. *American Behavioral Scientist* 43 (3): 377–391.

Star, Susan Leigh. 2002. Infrastructure and ethnographic practice: Working on the fringes. *Scandinavian Journal of Information Systems* 14 (2): 107–122.

Suchman, Lucy. 2002. Practice-based design of information systems: Notes from the hyperdeveloped world. *Information Society* 18:139–144.

Suchman, Lucy. 2007. *Human-Machine Reconfigurations: Plans and Situated Actions*, 2nd exp. ed. Cambridge: Cambridge University Press.

Turnbull, David. 2002. Performance and narrative, bodies and movement, in the construction of places and objects, spaces and knowledges: The case of the Maltese megaliths. *Theory, Culture, and Society* 19 (5–6): 125–143.

Wajcman, Judy. 1991. *Feminism Confronts Technology.* Cambridge: Polity Press.

Watts, Laura. 2008. The future is boring: Stories from the landscapes of the mobile telecoms industry. *Twenty-First Century Society: Journal of the Academy of Social Sciences* 3 (2): 187–198.

Watts, Laura. 2012. Orkneylab: An archipelago experiment in futures. In *Imagining Landscapes: Past, Present, and Future*, ed. Tim Ingold and Monica Janowski, 59–76. Farnham: Ashgate.

# 3 Freifunk: When Technology and Politics Assemble into Subversion

Gregers Petersen

Freifunk is an assemblage of a friction-filled multiplicity of interests and effects of technology use in everyday life in the twenty-first century. It is a community that originates in Berlin, a particular form of social movement, and a specific technological approach to computer networking. The tale of Freifunk's emergence and distillation tells of a process in which the everyday tactics of solving one's own problems (in this case, a lack of Internet access) are integrated with a more general strategy of political subversion. Thus it is also a tale of how technical appropriation becomes a sociotechnical subversion. To translate the word *Freifunk* into English: *frei* means "free," and *funk* is best understood as "radio waves" or the act of transmitting radio waves. Alternatively, Freifunk could be read as meaning "free broadcast(ing)."

The practical aspect of Freifunk is a technically novel approach to building large-scale wireless networks based on the Optimized Link State Routing (OLSR) mesh protocol. It became possible to develop this approach through a set of creative materials for building and designing that consisted of free software (Linux), cheap wireless devices, and a particular networking protocol. Few things emerge out of nothing, and Freifunk was in part spawned by the work of the Chaos Computer Club (CCC). The CCC has, since its birth in the early 1980s, manifested a strong technoculture of hacking, modifying, and critical politics. Club facilities are spread across Germany and bring together free-software coders, radio amateurs, hardware builders, and political activists. However, other proximate causes caused Freifunk to emerge, including the particular relations of people, things, and the social and political context in Berlin during the first decade of the second millennium. A final (and essential) element is the subversive power of free software, building on its unrestricted circulation and multiple ownership (Galloway 2004; Strathern 2005; Leach 2005; Gordon 2008). Boltanski and Thevenot (2006) argue that all political involvement is in the end driven by principles of justice employed to evaluate whether a social situation is acceptable. Freifunk is an example of how technology and a politics of design can work together when people face and wish to react to what they experience as injustice.

On a general level, it is necessary in understanding social processes to pay attention to what people are doing with objects and things, and how meaning emerges through appropriation and re-creation (Miller and Slater 2000; Michaels 1994). To understand contemporary European society, it is necessary to pay direct attention to how and why technology enters the lived-in reality (Lievrouw 2006), and why it is adopted or rejected. Technological choices reflect processes of cultural transformation and innovation (Lemonnier [1993] 2002). In this case, my work leads to the exploration of how people find their own uses for new technological objects, how people explore ideas and find hidden potentials in off-the-shelf commodity technology. My discussion here shows how the choice of technology is both political action and a process of constant re-creation of society. The chapter moves through the lines and threads that weave together wireless technology and subversive politics. The geographic focal point is Berlin, Germany, in the present and near past. The subject is the subversion of wireless networking devices by a DIY (do-it-yourself) community. This includes the subsequent reality of spreading a new networking paradigm and the corporeality of a free alternative mode for digital infrastructure—in short, how technical creativity and innovation become subversion, the inherent mangling of original intentions (Pickering 1995), thereby energizing new effects.

The chapter opens with a sketch of Freifunk as empirical phenomenon, based on my personal involvement, and through this sketch follows the intertwined lines between the struggles of everyday life and of interpretation. Wireless technology, along with the question of why it is so interesting in this context, does need a more detailed elaboration. This then leads to a debate in which I unfold a set of generalized theoretical aspects, thereby facilitating a merging of the topics of technology and subversion into a more unified construct. Finally, I point to some of the more general questions being raised by the development of the Freifunk movement, and how the acts of these people become a conduit to new cultural and technological expressions. This includes addressing the way that "subversive" and "politics" are to be understood in the present world of friction (Tsing 2005) and global assemblages (Ong 2005).

## On a Roof

I have been involved in Freifunk since its beginning in 2003. At times my involvement has been close, at times more distant. My interest in Freifunk has from the onset been driven by my combined interest in wireless networking devices, free software (Linux), and friendship with like-minded individuals. My engagement has always been as an activist and direct participant (Nash 2001; Graeber 2009). It is my experience that understanding a cultural setting such as Freifunk requires both "being" and "doing" Freifunk. Freifunk in Berlin grew from almost nothing in 2002 to a mesh cloud comprising eight hundred to one thousand nodes by 2009. This cloud covers approximately one-tenth of the city, which means that, theoretically, 350,000 citizens can connect to the network.

My interest in wireless networking began in the early years of the century. Looking back, I remember asking myself, "What's that square plastic box hanging on the wall?" The next time I saw the IT department person standing on a rickety chair, hands stretched far above his head, fumbling with cables, unable to find the much-needed reset button, I asked him, "What is that thing?" A few curses left his mouth before he spat out the answer: "It's a wireless access point!" I decided that it was probably better not to explore the implications of this information at that moment. I waited patiently until the problematic situation was over, and then phoned an IT department friend and asked what this thing with the wireless access point was all about. At the time, I was employed as a research assistant, combined with being system administrator of a little flock of Apple computers. None of the computers I had under my wing were equipped with wireless networking interfaces. I was wondering why the university was building a wireless network.

Today, years and experiences later, it makes sense. By now the small—and often colorful—plastic boxes with their insectlike antennae are found everywhere. People take for granted that their laptops and entertainment devices, by definition, are connected to a wireless network. Everyone is always online and has free and ready access to information and fun; individuals and groups across the globe are living increasing parts of their life "connected." Something must have changed, then, since my initial investigation.

Following the assumption that a relationship between technology and society exists, that their relationship is creative and dynamic, and adhering to its own cultural logics, it is evident that wireless technology has added or changed something, as well as just becoming more ubiquitous. The existence of a relationship between technology and forms of social organization (Wiebe and Law 2000; Lemonnier [1993] 2002; Akrich 1993) can twist ethnographic inquiry into new (or old) directions. Scholars of material culture are again scrutinizing things and objects such as technological devices

(Myers 2001; Miller 2005; Henere, Holbraad, and Wastell 2007), looking at how they can present themselves as material with potential(s) that unfold through processual relations. As stated by the OpenWrt.org project, which produces the software platform for Freifunk, "For users this means the ability for full customization, to use the device in ways never envisioned."

A good way to find out what Freifunk "looks like" is to go to a rooftop in Eastern Berlin and, using a laptop, scan for visible wireless networks. If one finds a network that does not behave as expected (in technical terms) with the extended service set identification (ESSID) "Freifunk," it is possible to estimate the physical position of a wireless router somewhere in the vicinity that works with the Freifunk software.[1] When the correct roof and location have been found by using this method, it is not uncommon to find a setup that consists of an old plastic canister with a hole cut into the bottom. Inside the canister is a wireless router, sometimes in its original housing, at other times just the board, out of the top of which comes an antenna, glued and patched to ensure a watertight connection. The antenna might be round beam or directional depending on the distance to other nodes in the wireless mesh network. The plastic canister arrangement is commonly attached to a stake or piece of tubing with a mixture of plastic strips and gaffer tape. Different electrical and network cables run down and away from the canister, leading to the local owner and maintainer. It would be possible to follow the cables down and ring a doorbell. Alternatively, one could access the Freifunk Web site and find the public information about that specific mesh node. Like the technical details about the user software and how to modify it, all information about the Freifunk network in Berlin is publicly accessible. The core of this dynamic mesh cloud is transparency.

**Reunited Berlin**

The last decade of the past century generated an immense development in Berlin, a metropolis filled with empty spaces and room for new uses. Nobody really knew who owned what in the central and eastern sectors of the city, and squatters, artists, and entrepreneurs of all kinds grabbed the opportunity and moved in. The German government created a vision of Berlin reemerging as a technological and financial hot spot. An essential requirement was a "future-proof" digital infrastructure. The liberal political spirit followed a privatization strategy, assuming that private entrepreneurship and market forces were the only ways to meet the goal of digital infrastructure. Development rights and licenses were sold at public auctions in which privately owned telecom and IT companies and speculators challenged each other with bids easily reaching billions of euros. A close-knit weave of fiber-optic cables went into the ground of the former German Democratic Republic capital.

At the same time, artists brought new styles and activities to the area. Whole streets filled with squatted houses, and "the underground" became a spirited expression of a possible future (Matus 2001). With the century running out, the Y2K crisis and the bursting of the IT bubble interrupted the influx of capital and energy. Dust settled, unemployment grew, investments collapsed, and a sense of wasteland, or the loss of a frontier, made investors move elsewhere, leaving both old and new inhabitants to fend for themselves to some degree. People suddenly had to find their own solutions to problems formerly solved by either the state or the market. One problem was that large geographic areas of Berlin, specifically central and eastern, had no working solution for offering private citizens access to an Internet connection. All the high-tech fibers in the ground stayed "black" (a fiber-optic cable is based on conveying light), owing to either bankruptcy or lack of capital to fund the necessary last mile of infrastructure to actual dwellings and offices. The telecom sector looked the other way, focusing on extending existing business and waiting until times became better. A strange situation emerged in which Berlin was at one and the same moment a center of the Western world, with all its facilities and services, and a Third World region without sewers or Internet infrastructure.

One of my colleagues in Freifunk offered me a far longer version of the preceding tale. But it is only a prelude to the real story, a real, everyday story about how a group of people took matters into their own hands and eventually created a new paradigm for how to build an alternative and free digital infrastructure. It was based on cheap wireless network devices and unrestricted sharing of information. The early initiative grew, transformed, and became Freifunk.net. Freifunk embeds in itself further levels of meaning and reflects community values centered on autonomy, horizontalism, and strong inspirations from the free software world. All of this melded into the concrete development of a large-scale wireless network solution based on mesh/ad hoc routing between independent nodes, in other words, a network without central control.

Freifunk—as organization, community, and technological platform—began with the founding of a free nonprofit association at the end of 2002. A small group of people living mainly in the Friedrichshain area (just east of Berlin Mitte) witnessed how private groups in cities like Seattle and London had begun to build far-reaching wireless networks. First the members of the association tried to convince commercial stakeholders to enter a partnership, but none of the companies saw a viable business plan. Contacting state and public institutions yielded similarly negative results. The growing community of people who lacked and wanted Internet connectivity had to do it themselves.

**Networking Basics**

The tale of how Freifunk computer networking itself came into being is embedded in various interpretations of historical events (and rumors). This is the internal work of everyday organizational struggle and cooperation. I do not intend to explore this further here as a sociopolitical narrative. Instead I think it more pertinent to explore why wireless networking technology is such an enticing and "wild" thing, with so much persuasive power (Attfield 2000) when it comes to be converted, as hardware and protocol, to a shared political vision.

The networking of computers is, at least in the extended Western world, a fact of life. The physical layout is commonly created by and through the use of network cables (Ethernet is the de facto standard), and the actual exchange of data, or information, is based on connections between a server and client(s) (i.e., "managed mode," as it is known by its technical term). This specific way of managing a network of computers is extremely hierarchical. All the activities of the network depend on a centralized server, of which every other device acts as a client. Each client has to obtain an IP address from a local server (an Internet protocol address, a specific set of numbers within a predefined range). The server usually issues the IP address based on a request for the hardware MAC address (media access control address) of the client, which in human-readable media translates into six groups of two hexadecimal digits. When the connection has been established and confirmed—often through the use of a password authentication system—then, and only then, will the server route data and requests to and from the client across the local network and beyond. In practical terms, the client devices are at the bottom of the hierarchy. The client has no influence on either the services offered (shared drives, printers, etc.) or the actual routing of traffic. When it comes to the integration of *wireless* networking technology into the managed-mode model, with its inherent metaphors and assumptions, this form of server and client is directly replicated. The infrastructure simply becomes "wired" without cables. One might then ask: why is wireless technology so interesting, and why does it contain so much potential and wild attitude hidden under a layer of assumptions about use and trajectory? Galloway (2004, xvii) has discussed the significance of networking protocols and in particular the embedded political economy expressed in terms of control and institutionalization. A new protocol, including new rules for the connections, changes the game and a offers a different trajectory.

A cursory exploration of the various wireless settings and options on a laptop computer today, combined with a sprinkle of reflection on why antennae seem to be such an important aspect in present-day computing, can be a revealing experience. Wireless settings commonly offer two choices, either "connect to a wireless network" or "create a computer-to-computer network" (ad hoc). The first option is the default managed mode; the second choice tells a story of a different kind. It indicates that the central server is not always required. It is possible to link two or more computers or devices together in an alternative mode of "nonmanaged" network.

A wireless network interface is a little two-way radio, and radios are made to connect with other radios. Ad hoc mode is a network connection method mainly associated with wireless network technology, but it works as well on a standard wired network. The connection is established for the duration of one session and requires no central access point, or server. Instead devices discover other nodes within range to form a network for those same computers. A wireless ad hoc network is thus a decentralized wireless network, with nodes instead of clients. The network is ad hoc because each node is willing to forward or route data for other nodes, so that the determination of which nodes forward data is made dynamically based on the network connectivity. This contrasts with wired networks in which servers or routers perform the task of routing. It also contrasts with managed wireless networks, in which a special node, known as an access point or wireless router, manages communication among the connected nodes.

Ad hoc networks are commonly grouped together into the category of "mesh" networks, based on different routing protocols (OLSR, etc.). A mesh network consists of individual nodes that search for target nodes and provide the layout of the network itself by flooding the network with broadcasts that are forwarded by each of the other nodes. Flooding a network with broadcast "packages" can seem problematic, but the traffic generated is similar to the traffic a normal server generates. A mesh network is reliable and offers redundancy. Redundancy offers a robust level of service, and this is one of the distinct and desirable qualities of mesh networking. When one node can no longer operate, the rest of the nodes can still communicate with each other, either directly or through one or more intermediate nodes. The individual node at intervals calculates new dynamic routing tables (a map of the network), and combined with specialized protocols, this map provides stable connections, even if nodes are mobile. A wireless ad hoc network adapts itself to the existing conditions and dynamically changes over time. It easily handles coming and going of nodes. Mesh clients can be laptops, cell phones, or other wireless devices, while the mesh routers themselves are specially configured as radio nodes and forward traffic to and from the gateway(s). Gateways may be (but are not required to be) connected to the Internet. The coverage area of the radio nodes working as a single network is sometimes called a mesh cloud. Access to this mesh cloud depends on the radio nodes working in harmony with each other to create a radio network, and this requires a common routing protocol and frequency spectrum.

A mesh network is by definition horizontal, nonhierarchical, and self-organizing, based on the concept of a node being simultaneously client and server with a routing protocol that always seeks the best way. Wireless networks reside in the 2.4 GHz and 5 GHz frequency range of the spectrum (also known as the microwave range). These two "narrow" bands of radio waves have internationally been declared deregulated spectrum. In other words, with a few exceptions, everyone is free to use these slices of spectrum as he or she pleases.

**Freifunk**

In the past, a physical wall forced Berlin apart. At the time of the early Freifunk development, the geographic divide was digital. Among the residents of east Berlin were Freifunk members who had long-term amateur radio experience, Linux hackers, hardware engineers, artists, and squatter activists, and all agreed from the outset that the solution to their need for sharing information and accessing the Internet was wireless. Small-scale experiments with building wireless links spanning up to several kilometers and bringing connectivity to individual buildings were already known about. However, at the time, all other free networks had implemented normal managed networks, and in the beginning Freifunk followed this well-paved path. A grand plan was created, bringing together a bricolage of groups, entities, and cultural and art institutions to build a large "Berlin BackBone," spanning the Berlin Mitte. The BackBone was to be free for people and groups to connect to, attach digital branches, and extend the coverage. But the plan never became a functional reality. The idea of a common and free backbone killed itself in fights over control and resources. (Who was to pay for the central equipment? How to collect payments for bandwidth? Who should and should not manage the infrastructure? Could individuals or groups deny others access? etc.) The quarrels proliferated, and the grand backbone vision was never realized. One of the explanations I gathered about the project's quick decline emphasized that "everything was drowning in people who wanted to enforce hierarchy and manage what other people should do—or had money, and thought they could buy influence. Nobody was interested in doing all the hard work, and then just accept that the network was taken away by other people. We don't do things in that way, and had to find a different solution."

For a while, the Freifunk community focused on weekly "learning-by-doing" meetings, modifying cheap wireless routers and experimenting with homegrown antennae. Antennae are an art form in themselves, but they can be made from simple objects such as old metal cans or pieces of wire netting. Wireless routers could be bought almost on every street corner, and many of the available models used Linux as their operating system, meaning that the manufacturer had to comply with the General Public License agreements that enforce the release of human-readable versions of modified software code. It became routine among Freifunkers to "reflash" (reconfigure) a modified Linux system (firmware) onto the wireless routers they collected, with a program developed locally out of Linux building blocks called OpenWrt being the preferred distribution. Line-of-sight and directional antennae are typically needed to establish stable connections between distant points. Numerous small and localized networks began spreading, but linking them together was difficult owing to factors such as line-of-sight interruptions and social dissonance.

Then, during the spring of 2004, something happened. The first experiments with the OLSR mesh protocol took place. Informants mainly agreed that "a couple of people had found that there was ample room in the memory of a Linksys WRT54G wireless router to accommodate the OLSR daemon, and played with the configuration of a system for a while." A few months later, the first small mesh cloud emerged in the district of Friedrichshain, which was already known by the alter ego WlanHain.[2] From this moment, the Freifunk network quickly merged into a stable form. A set of components, or building blocks, was developed and made available to the public (freifunk.net):

- Freifunk firmware, an OpenWrt-based replacement (Linux) for the software provided on off-the-shelf wireless routers
- A suitable wireless router, initially the Linksys WRT54G
- Homemade antenna for outdoor mounting

The firmware was and is free to download and comes with instructions for use and configuration. The required wireless devices were readily available at stores, as cheap off-the-shelf commodities (40–50 euros). Building a directional tin-can antenna, a "cantenna," costs approximately 10–15 euros in materials and requires only the drilling of a single hole and a minute amount of soldering. As soon as the idea was brought to life, the autonomous Freifunk network and its nodes began multiplying.

**Freifunkerei**

To become a user of the Freifunk network, an individual must acquire a wireless router, "flash" it with the Freifunk firmware, and finally add a simple configuration to the device. The practical process often takes place at one of the weekly soldering meetings. In Berlin these are held at C-base, an imaginary old spaceship turned cultural center in the district of Mitte. A person interested in joining or extending the network can attend a meeting and receive the necessary help (depending on individual skill level). The mesh network created by all the active Freifunkers is an open and public network. This network allows for the free—free as in no charge, as well as open—exchange of data and services (instant messaging, P2P file sharing, VoIP protocols, etc.).

The economic model is based on cooperation, sharing, and mutuality rather than an individual commodity fixation. The Freifunk network exists under the terms of the Pico Peering Agreement (PPA). The PPA is a formalized structure that provides "the minimum baseline template for a peering agreement between owners of individual network nodes," in which "owners of network nodes assert their right of ownership by declaring their willingness to donate the free exchange of data across their networks." The agreement stipulates free transit, open communication, no guaranteed level of service, acceptable use, and local amendments. This contrasts with commercial and corporate models of Internet and information access.

One could claim that it is the sole interest of commercial companies to produce technology in a form that is static and unmodifiable, to maintain a product as an unchangeable commodity artifact. The practice and development underlying Freifunkerei are an assemblage of material and the creation of knowledge in social process—an ongoing dialogue between what I hope I have now demonstrated are technologically "wild things" and particular people. The Freifunk individuals and community that explored and pushed the development of large-scale mesh networking—something previously found only in university and military research experiments—represent what one might call an emergent set of cultural traditions understood as the process of sociotechnical reproduction and the passing on of norms, values, and institutionalized practice.

## Traditions

Freifunk as social movement and community is in its assembled form a recent expression of more general social, political, and technological change. As movement, Freifunk consists of individuals who engage in collective action. They are linked by dense but often geographically dispersed networks, and they are given solidarity by their conflictual relations with clear opponents (Della Porta and Diani 2006). The most visible friction is displayed in spheres involving capitalist market forces and hegemonic state structures with their "denial of service." The Freifunkerei have their foothold among squatters, software hackers, autonomous left groupings (Autonome Linke), and other groups of the margins and perimeter. There are significant anarchist currents centered on individual autonomy and horizontal consensus-based decision making. One might draw parallels between Freifunk and Pierre Clastres's (1989) interpretation of society against the state and other studies of horizontal direct-action communities and movements (Sitrin 2006; Graeber 2009). A world turned upside down is an ongoing theme, and the Freifunk mesh network solution reflects this. A wireless mesh network is everything that the normal managed network is not supposed to be. There is no fixed routing, nodes come and go, and there is no hierarchical server. Maintaining this technical position is like other cultural technological elements reproduced in particular ways based on local traditions.

I was spending recurring periods of time in Berlin during the first explosive growth of the mesh cloud. *Wave-loeden* (wave soldering) was one of the weekly events drawing people together. Active members used it as social gathering, with space and time for hacking and modifying. These events also provided an opportunity for individuals who wanted to join Freifunk by setting up a new mesh node—and thereby extending the network—to receive practical help with reflashing and configuration.

On one of these evenings, a young man with a shiny new Mac laptop brought a wireless router still packed in its original plastic wrap. He asked for help with adding a mesh node to the network. First he was told to unpack the router, use his laptop to download the required Freifunk firmware, and register an IP address for his device. (IP addresses are related to the geographic locality of the mesh node and are needed to finish the simple configuration. The individual devices in the network all need to be configured with addresses within the same IP range.) There was some interest in the detail that he wanted to place the node in a more or less empty area along the Spree River, but it was also understood that this would be a positive extension of the mesh cloud. Hands helped with booting the router into rescue mode, connecting a network cable to the laptop, and showing him how to use the necessary command line applications, as he had none of the required skills or experience. Then someone asked why he had chosen the appointed locality for the node. It emerged that he was building a beach bar with some of his friends, and they wanted to offer their customers free Wi-Fi. Now, normally bar, beer, and beach have positive connotations. The questioner then asked what he intended to do regarding the choice of Internet gateway. The young man's answer pulled smiles off several faces. His plan was to grab and use the Freifunk network (and bandwidth) and make money from customers who paid for drinks and food. He did not intend to add an Internet gateway to the mesh cloud at some other point; the bar was a summer project intended to last from three to four months. Suddenly the hands were no longer there, and he was surrounded by a barrier of empty space, holding on to a nonusable wireless router. It took him a little time to realize that he was on his own. I am not sure if he ever came back. What had been displayed was a simple reaction to a person who publicly acknowledged that he would not fulfill the obligation of mutuality and was seeking personal profit. Now, no one in Freifunk would deny that money is needed to feed the Freifunk network. Money is used to pay for the numerous ADSL connections that function as Internet gateways, equipment has to be bought, electricity is a prerequisite, and so on. These costs are covered by individuals, or sometimes small groups pool together. The point is that nobody can claim or make money from others for accessing the network. People and mesh nodes are all individual, but producing and maintaining mutual relations create a network, a horizontal society through a technology that has been subverted to meet cultural values.

**Trajectories**

Consulting a dictionary will reveal that the term "subversion" translates to a systematic attempt to overthrow or undermine a government or political system by persons working secretly from within. In continuation, "subversive activity" is defined as the lending of aid, comfort, and moral support to individuals, groups, or organizations that advocate the overthrow of incumbent governments by force and violence. I always feel slightly overwhelmed by the grandiosity of these definitions. The writing I produce here is not about a well-planned violent overthrow of a political regime but in fact much more about the exploring of alternative technological trajectories. I suggest that these are in themselves an act of everyday resistance against consciously created limitations and unquestioned assumptions. The expression of subversion, in this context, is about daily tactics, not about well-planned strategies. It is not possible to ignore the fact that the activities I aim to describe are political actions, intentionally and deeply subversive. Intention is a central element in my approach to subversion. One might ask why this is the case. David Graeber has explored political action and, in a summary form, explains: "As a minimal definition, political action is action meant to influence others who are not physically present when the action is being done. This is not to say it can't be intended to influence people who are physically present; it is to say its effects are not limited to that. It is action that is meant to be recounted, narrated, or in some other way represented to other people afterwards; or anyway, it is political in so far as it is" (2007, 130).

How people find their own use for new technology, how choices are enacted, is inherently political. The introduction of a particular technology might have an intended outcome, but people manifest their own intentions and deliberately, or sometimes just by chance, enter the spaces of power through what in the end become acts of subversion. It is the tactics of many small steps, as discussed by Colin Ward (1982), in this case how the exploration of technological potentials becomes a combined autonomous act. Such diverse acts of use and reuse of technology require a focus on the social dimensions of the use, circulation, and appropriation of technology among communities and social movements. They require an awareness of why certain technologies become excluded, fail, or are subverted as a way of actively protecting culture and traditions. Processes of resistance during everyday life in the present, as de Certeau (1984) has explored, often involve the use, reinterpretation, and subversion of technology, and we must recognize these acts as inherently political. It is important to think about the ways in which professional, political, and economic factors shape and form the technology we surround ourselves with. Technologies are not purely technological; "instead, we have said that they are heterogeneous, that artifacts embody trade-offs and compromises. In particular, they embody social, political, psychological, economic, and professional commitments, skills, prejudices, possibilities, and constraints" (Bijker and Law 2000, 7).

## The Subversive

Technological devices are meshed into everyday life, and as Lefebvre (1987) has noted, the everydayness of life is likewise the most unique and universal condition; it is both social and individual. The most obvious is often the best hidden. Technology is likewise nonspecial and mundane; in some ways, it disappears in the physicality of daily activities. But technology in itself is a visible expression of this everydayness. Objects and "things" can, as Judy Attfield (2000) states, "be defined much more broadly as 'things with attitude'—created with a specific end in view—whether to fulfill a particular task, to make a statement, to objectify moral values, to express individual or group identity, to denote status or demonstrate technological prowess, to exercise social control or to flaunt political power" (12).

Wireless technology presented itself with the potential "attitude" and a "wildness" to Freifunkerei who had been disenfranchised by both state and market. I believe it is a common assumption that technology represents a certain order or is set within a field of containment. It is assumed that the developers (or producers) of a given technological product are able to predict and control how the product will be used within a predefined place and space. And justifying this, ordinary users would probably never get anything done if they had to think about every single "thing" and act as possessing attitude and alternative potential all the time. The conduct of everyday life somehow demands a tactical lack of curiosity, according to Bijker and Law (2000). Or does it? Society itself is built and reproduced along with things, and technological things do not always follow expected trajectories. This raises questions about agency and subject–object relations when it comes to the emergence or introduction of new technology. Looking at technology and how it meshes into everyday life can motivate processes of appropriation and reinterpretation. Control or lack of control over technological trajectories is a succulent direction to explore.

Andrew Feenberg (n.d.) introduced two principles that put the intersection between technology and society into perspective and highlight its ambivalence. Feenberg observes: "If authoritarian social hierarchy is not technically necessary, then there must be other ways of rationalizing society that democratize rather than centralize control." He argues that it is in fact the intention of emerging social movements to change technology, and he shows how new kinds of actors breach the assumed divide between experts and normal people.

The first principle is the conservation of hierarchy, and it explains the extraordinary continuity of power in advanced capitalist societies over the generations. It is a continuity made possible by technocratic strategies of modernization, despite enormous technical changes. The second principle is subversive rationalization, by which Feenberg understands the technical initiatives that sometimes accompany the strategies of structural reform pursued by union, environmental, and other social movements. To undermine requires both tactics and strategy.

Tactics are intrinsic to the implementation of strategies, and there are a thousand ways of playing and outplaying the games of the other. Michel de Certeau (1984) explains the tension between strategies and tactics by the multiplicity of codes coexisting in any society. Some of these establish themselves hegemonically, while others remain marginal and exist only in the special uses they determine. The technical code of society is a syntax that is subject to unintended uses and may subvert the framework of choices it determines. Strategies expose the substantive implications of the apparently neutral technical management of modern organizations, and a focus on and analysis of tactics bring out the inherent limits of dystopian rationalization.

**What Did Freifunk Do?**

To me, de Certeau suggests a way of understanding resistance as neither individual moral opposition nor just another policy. Resistance, as a general modification to which strategies are subject, belongs to another order. Just as operational autonomy serves as the structural basis of domination, so a different type of autonomy is won by the dominated, an autonomy that works with the play in the system to redefine and modify its forms, rhythms, and purposes. In the case of Freifunk, the play between strategy and tactics resulted in unauthorized technical improvisation, reappropriation, and the practical innovation of a new networking paradigm.

Technical protocols and solutions can be ways of controlling social structures (Galloway 2004). The disenfranchised individuals living in Berlin in the second decade of the reunification either already knew or learned this by doing. Freifunk individuals and community took a radical approach to solving a problem normally encountered in Third World settings. The German state had abandoned the infrastructure needs of its citizens, leaving the problem to be solved by market forces. So why should people not be able to build their own network solution and tie it together in their own ways? When Freifunk began, there was overwhelming friction between available technical models, forms of social organization, and cultural values. What took place was a transvaluation of the technology itself: "Transvaluation situates itself in a radical manner on the edge of time, and only here. Transvaluation is the productive event" (Negri 2005, 218).

Bringing technology in sync with the local society in this case happened through a productive and creative event. Society itself is built and reproduced through the temporal unfolding of relations between people and things, and technological things do not need to follow planned trajectories. Freifunk did not assemble out of nothing; the material was already in flow and freely available (Petersen 2008): open-source software, cheap and powerful hardware devices, and experience with DIY actions. Linking the subversive politics of squatting a building to the practical appropriation and subversion of Internet infrastructure was not a great leap.

Horizontalism and the return of models of multiple ownership (Leach 2005; Strathern 2004, 2005) and the inherent free flow and circulation of information are together questioning the technological top-down model of capitalism—and can be seen to subvert the current structures of the hegemonic (techno)society. Freifunk is an example of how simple things such as wireless technology can leave a fixed trajectory. Instead of being a victim of conscious acts of control and the oppression of underlying assumptions, the Freifunk community expanded a politics of subversion.

### Afterword

Freifunk did expand rapidly beyond the confines of Berlin. Today it has become a global phenomenon. Freifunk-style mesh networks are emerging in many places, from other German cities to Himalayan valleys, from Africa to linking villages and favelas in Latin America. The largest Freifunk-inspired mesh cloud has emerged in Athens, Greece, in recent months. This can only be a source for interesting thoughts as the world watches the implosion of the Greek state, economic system, and political reality. Trajectories are never given.

### Acknowledgments

I would like to thank Julian, Jürgen, Alex, Cven, Horst, Tomas, Sebastian, Elektra, Steffi, Tilo, Kloschi, Saul, James, John, Felix, Jo, and all the other Freifunk and OpenWrt friends who have given me so much.

### Notes

1. The extended service set identification (ESSID) is the name broadcast by a given wireless network.

2. "Wireless local area network–hain."

## References

Akrich, Madeleine. [1993] 2002. A Gazogene in Costa Rica. In *Technological Choices*, ed. P. Lemmonier. London: Routledge.

Attfield, Judy. 2000. *Wild Things: The Material Culture of Everyday Life*. Oxford: Berg.

Bey, Hakim. n.d. Permanent TAZs. http://www.hermetic.com/bey/paz.html.

Bijker, Wiebe E. 2000. The social construction of fluorescent lighting, or How an artifact was invented in its diffusion stage. In *Shaping Technology/Building Society*, ed. W. E. Bijker and J. Law. Cambridge, MA: MIT Press.

Bijker, Wiebe E., and John Law. 2000. General introduction. In *Shaping Technology/Building Society*, ed. W. E. Bijker and J. Law. Cambridge, MA: MIT Press.

Boltanski, Luc, and Laurent Thevenot. 2006. *On Justification: Economics of Worth*. Princeton, NJ: Princeton University Press.

Clastres, Pierre. 1989. *Society against the State*. New York: Zone Books.

de Certeau, Michel. 1984. *The Practice of Everyday Life*. Berkeley: University of California Press.

Della Porta, Donatella, and Mario Diani. 2006. *Social Movements*. Oxford: Blackwell.

Feenberg, Andrew. n.d. Escaping the iron cage, or Subversive rationalization and democratic theory. http://www.rohan.sdsu.edu/faculty/feenberg/schom1.htm.

Galloway, Alexander R. 2004. *Protocol*. Cambridge, MA: MIT Press.

Gordon, Uri. 2008. *Anarchy Alive*. London: Pluto Press.

Graeber, David. 2007. *Lost People*. Bloomington: Indiana University Press.

Graeber, David. 2009. *Direct Action: An Ethnography*. Oakland: AK Press.

Henare, Amiria, Martin Holbraad, and Sari Wastell. 2007. Introduction. In *Thinking through Things: Theorising Artefacts Ethnographically*. London: Routledge.

Leach, J. 2004. *Creative Land: Place and Procreation on the Rai Coast of Papua New Guinea*. Oxford: Berghahn Books.

Leach, J. 2005. Modes of creativity and the register of ownership. In *Code*, ed. Rishab A. Ghosh. Cambridge, MA: MIT Press.

Lefebvre, Henri. 1987. The everyday and everydayness. *Yale French Studies* 73:7–11.

Lievrouw, L. A. 2006. New media design and development: Diffusion of innovations vs. the social shaping of technology. In *Handbook of New Media*, ed. L. A. Lievrouw and S. Livingstone. London: Sage.

Lemonnier, Pierre. [1993] 2002. Introduction. In *Technological Choices: Transformations in Material Cultures since the Neolithic*, ed. P. Lemonnier. London: Routledge.

Matus, Victor. 2001. The once and future Berlin. http://www.hoover.org/publications/policy-review/article/6707.

Michaels, Eric. 1994. *Bad Aboriginal Art*. Minneapolis: University of Minnesota Press.

Miller, Daniel. 2005. Materiality: An introduction. In *Materiality*, ed. D. Miller. London: Duke University Press.

Miller, Daniel, and Don Slater. 2000. *The Internet: An Ethnographic Approach*. Oxford: Berg.

Myers, Fred R. 2001. Introduction. In *The Empire of Things*, ed. F. R. Myers. Santa Fe, NM: School of American Research Press.

Nash, June. 2001. *Mayan Visions: The Quest for Autonomy in an Age of Globalization*. London: Routledge.

Negri, Antonio. 2005. *Time for Revolution*. London: Continuum Books.

Ong, Aihwa. 2005. Ecologies of expertise: Assembling flows, managing citizenship. In *Global Assemblages*, ed. A. Ong and S. J. Collier. Oxford: Blackwell.

Petersen, Gregers. 2008. Circulating property. Paper presented at Rethinking Economic Anthropology Conference, SOAS and London School of Economics, January 11–12. http://www.rethinkingeconomies.org.uk/web/d/doc_81.pdf.

Pickering, Andrew. 1995. *The Mangle of Practice*. Chicago: University of Chicago Press.

Pico Peering Agreement. http://www.picopeer.net/PPA-en.html.

Sitrin, Marina. 2006. *Horizontalism: Voices of Popular Power in Argentina*. Oakland: AK Press.

Strathern, Marilyn. 2004. Introduction. In *Rationales of Ownership: Transactions and Claims to Ownership in Contemporary Papua New Guinea*, ed. L. Kalinoe and J. Leach. Herefordshire: Sean Kingston Publishing.

Strathern, Marilyn. 2005. Imagined collectivities and multiple ownership. In *Code*, ed. Rishab A. Ghosh. Cambridge, MA: MIT Press.

Tsing, Anna L. 2005. *Friction: An Ethnography of Global Connection*. Princeton: Princeton University Press.

Ward, Colin. 1982. *Anarchy in Action*. New York: Freedom Press.

# 4 Postcolonial Databasing? Subverting Old Appropriations, Developing New Associations

Helen Verran and Michael Christie

## Introduction

Databasing is a particular contemporary way of "doing knowledge" with information and communications technologies. Here we write out of our experience with what we understand as a postcolonial databasing project (see http://www.cdu.edu.au/centres/ik) that aimed to devise some specific forms of databasing that might be useful for Aboriginal Australian users. At the project's core was the understanding that in engaging with ICTs in thoughtful ways as a group, we were connecting up (and keeping separate) Aboriginal knowledge traditions and technoscientific traditions.

Originally proposed to examine and develop digital technologies for the intergenerational transmission of traditional Aboriginal knowledge, the project, which we called "Making Collective Memory with Computers," was slightly unstuck from the beginning. We found industry partners in the Australian Research Council project (the local herbarium and the local Aboriginal Land Council), which wanted to develop further their own databasing solutions (ethnobotany databases, resource mapping), and our Aboriginal coresearchers, who wanted to borrow (then buy) the project's cameras, computers, and expertise for the new opportunities they might afford. We were unsure at first how to proceed, and our work eventually amounted to following people around and watching the emerging uses of digital technologies as they borrowed cameras and cables, configured digital objects and their enabling software and hardware, juxtaposing, obscuring, revealing, working digital forms of ceremonial common ground as "*garmas*," producing local solutions to local problems. We learned to see the digital files not as containers of knowledge (through the conventional practice of representation) but as active participants, artifacts of previous knowledge-making episodes that were being enlisted and configured in extremely lively conversations. To capture this sense of tentative and emergent inquiry, our contribution to this volume takes the form of questions and answers. We consider puzzles that frequently emerge when Aboriginal Australian knowledge traditions engage in databasing projects and use of ICT.

**What Is Postcolonial Databasing?**

As analysis of using contemporary information and communications technologies (ICTs) in struggling to deal with and go beyond colonial archiving legacies, postcolonial databasing seeks to focus on specific practices and engagements, orderings and framings, in collective memory making and knowledge generation in non-Western contexts. We want to think about how local knowledge-making and archiving practices using ICTs operate within particular histories and logics, pasts that amounted to unthinking appropriation of "other" forms of knowledge. The materialities of colonial archiving practices are crucial in what could and could not be achieved in colonial archiving (for an overview, see Hamilton et al. 2002), but equally consequential are the working imaginaries within which they were and are embedded. Attending to our working imaginaries is central in postcolonial databasing.

Working imaginaries are narrative and image, metaphor and analogy. They frame and explain; they are stories and pictures that figure, prefigure, and refigure relations. As such they indicate that working knowledge traditions might be interrupted and subverted, leading to slowing down (Stengers 2005). With slowing down, working imaginaries emerge more clearly as metaphysical commitments and the means of articulating such commitments. The unique assumptions that lie at the core of all knowledge traditions are felt as both limits and possibilities, and with that comes a chance for developing futures different from pasts. Importantly, working imaginaries of archiving in general and databasing in particular are institutionally embedded; they are collective practices. Postcolonial databasing, then, is engaging newly emergent digitizing technologies within a working imaginary that reflexively seeks to promote a postcolonial moment on the ground, going on together, in the here and now (see Verran 1998, 2002).

The idea of postcolonial databasing came to the fore at the very start of our project, when in an audit we found more than one hundred databases of "Aboriginal ecological knowledge" located in official institutions in the top end of Australia (Scott 2004). Yet we found no evidence of Aboriginal people either using official collections that represented their natural knowledge traditions or developing their own collections of digital items until we started working with them on their own land, following them around as we all worked together with cameras, sound recorders, and computers. Even then, these Aboriginal knowledge workers who were developing their own private collections of digital objects were challenged by the traces of Western metaphysics apparently *inside* the software they had taken up. Generalizing from this experience, we claim that the coming to life of digital databases within a postcolonial moment involves recognizing the working together of disparate knowledge traditions.

In postcolonial databasing, we are struggling to find ways for limited and useful connectings, and also to ensure separations and distinctions. We need to take care that one way of doing collective memory does not overwhelm the other. Aware of how easily other lifeways are crushed by modern ways of going on, as we research how ICTs might generatively be engaged in postcolonial contexts, researchers need to watch themselves. In other words, reflexivity, reflecting openly, assiduously, and continually on collective practices, is crucial. It's important to try out "doing databasing" with different groups, heeding their differing purposes, and in the research keep an eye out, to watch as a stranger might, just what is being done, and how. Postcolonial databasing needs a way to show both those inside and those outside how to keep enthusiasms under check. The Indigenous Knowledge and Resource Management in Northern Australia (IKRMNA) project might be regarded as an early exemplar of postcolonial databasing. In this question-and-answer essay, we refer to this project in particular and Yolngu Aboriginal Australian knowledge practices more generally to explore the idea of postcolonial databasing and how we might avoid old forms of appropriation and, through informed and reflexive practices, develop associations generative for both sides.

## What Aspects of Knowledge Practices Are Important to Aboriginal Australians in Developing Digital Databasing?

Many Aboriginal people in northern Australia are using digitizing technologies—computers, video and still cameras, audio recorders, manipulation and presentation software, and written texts—to generate digital items that can contribute to various forms of Aboriginal knowledge work. Just as in the sciences, so collective memory making in its various guises is important in using and making knowledge in Aboriginal knowledge traditions

(Bowker 2006). However, only recently have digital objects, and their status and location, become part of Aboriginal knowledge practices. Nevertheless Aboriginal people have been keeping audiocassettes of recordings of important stories and songs secret and hidden in sacred woven dilly bags for some time now, along with more traditional items involved in traditional knowledge practices. A lively politics of sharing and concealment governs these objects. This politics of revealing and hiding also informs digital databasing.

When it comes to actually doing the work of assembling a collection of digital objects that might be useful in an Aboriginal knowledge context, several aspects of how to do it immediately become problematic. For example, what sorts of things digital objects *are* in an Aboriginal context of knowing turns out to be surprisingly puzzling and unpredictable. It is difficult to tell the ontological status of those objects. They represent, but they also perform. Unlike Western institutions, Yolngu institutions defer the separation between the ontic and the epistemic in their knowledge practices. They eschew foreclosure that might precipitously convert digital objects to less ambiguously epistemic objects. But as we have seen, in achieving this, Aboriginal Australian elders struggle against the grain of digital technologies designed to represent, and use them with seeming ease in their traditional practices, where knowledge is always actively performative rather than representational. And more and more they cannily exploit the possibilities that digital objects reveal in achieving political ends in dealing with representatives of mainstream Australia (Verran and Christie 2007).

However, when Western institutions, even those like land councils and libraries promoting Aboriginal interests, set up community-level archives and knowledge centers, just as in the more official databases we found in our audit, the performative functions of digital objects become obliterated or marginalized. Connections with their performance by their makers get lost as those connections become pieces of metadata, the locatedness of the images in the land gets lost as objects from a range of sources find themselves together, and the power to conceal and reveal becomes compromised as community-level (rather than individual or family-level) archives are set up and supported by outside institutions.

Performativity and the politics of revealing and hiding are significant in considering how a database might be organized so that it could be useful to Aboriginal people as they do their knowledge in their own ways, using their own forms and structures. It is not at all clear in the beginning how to proceed. Needing to think through these issues helps us to understand how

# Postcolonial Databasing?

ICTs engage with knowledge traditions generally, and Aboriginal Australian and technoscientific knowledge traditions in particular.

## Why Use the Term "Knowledge Traditions" Rather Than "Knowledge Systems" When Discussing Postcolonial Databasing?

Both "systems" and "traditions" are metaphors, working images of how we understand knowledge using and knowledge making. "Tradition" comes from the Latin word *tradere*, "to give." "Traditions" emphasizes human communities "doing" their knowledge, giving both across generations and to other knowledge communities. "Systems" comes from the ancient Greek word *systema*, meaning "set." "Systems" implies a concern with boundaries and focuses on framings and separations. It emphasizes the structures of knowledge.

In using "traditions," we are not denying the importance of structure in knowledge. The practical difficulties that can arise in working disparate knowledge traditions together are often caused by differences in the ways things are framed and structured. In using "knowledge traditions," we mean to draw attention to the fact that all human communities have complex and varied ways of dealing with such issues in their practices of knowledge using and making. The ways that framings and reframings are managed when knowledge traditions work together is part of what is at stake in a postcolonial databasing. When we want to connect and separate knowledge traditions, we need to think about and discuss the various sorts of reframings we need to do to achieve our purposes while working in good faith both with our familiar knowledge practices and with those of the other—and, of course, this applies on both sides. Using the term "traditions" and remembering its origins in giving implies purposeful openness in "doing knowledge." Traditions can be reimagined and repurposed piece by piece, carrying the insight that knowledge is done in the here and now. In working a tradition, collective memory is understood as a present-oriented activity, and so are imagined future generations of learners.

The presentist focus that comes along with the term "tradition" rather than "system" also explains our odd use of the term "databasing"—not usually thought of as an activity and referred to with a verb. The verb reminds us that understanding the design and use of databases as alternative moments of a single activity is important, not least to understanding where accountability for design lies (Suchman 2002; Verran and Christie 2007).

## What Are the Terms We Can Use in Discussing the Various Sorts of Reframing in Postcolonial Databasing?

We are interested, first of all, in eschewing appropriation and reimagining generative associations between knowledge traditions in postcolonial databasing. In using and making knowledge, people are extremely aware of some framings. Other framings are deeply hidden; in postcolonial databasing, both forms of knowledge framing are important. For example, all knowledge traditions have experts in various fields and disciplines. Access to expert knowledge must be managed. Of course, a difference in scale exists between Aboriginal knowledge traditions and technoscientific traditions. Aboriginal knowledge communities are generally small—small enough for all those involved to be personally acquainted. But that difference in size does not alter the fact that there is "outer," "inner," and "secret" knowledge, and institutional ways of managing access to those levels. The institutional arrangements involved in using and making knowledge, not the scale, express theories about what knowledge is in any knowledge tradition.

The institutional arrangements and practices involved in working a knowledge tradition embody the ways that knowledge is justified as true. "Epistemic" is the general term we use to name this aspect of knowledge. Epistemology involves discussing and considering the management of these institutionalized structurings and reframings. These terms come from the Greek word for knowledge: *episteme*. The *-ology* bit of the term means "to study."

There are also divisions and definitions that knowledge users and makers are far less aware of. Becoming sensitive to this level of difference can be crucial in successfully working together disparate knowledge traditions. These structural differences are embedded in language use and in the ordinary generalizing we do when we use numbers. Here people are working at the level of assumption; things are usually just taken for granted as people go on together. In working disparate knowledge traditions together, people must bring these assumptions and what they take for granted out into the open. Often, especially in the beginning, that process is not comfortable. Philosophers name this profound level of framing the ontic level. Ontology, then, is the study of what there is, or more precisely what knowers are committed to there being. "Ontic" and "ontology" come from the ancient Greek term *onto-*, a form of the verb form *eimi*, or "am" in English, part of the verb "to be."

The point is that the ontic is *not* by its very nature different in kind from the epistemic; both refer to what we might call conventions about going on in the world. Distinctions between the two aspects of knowledge are accomplished in rhetorical practice, distinguishing the forms that involve justification from those that do not. Institutional edifices solidify and usually hide both the fact that distinctions between the ontic and the epistemic are collective achievements and how they are achieved.

So, taking possibilities for postcolonial databasing seriously, we have two distinct tasks. One is to collaborate on some of the ontic work that is being done in the particular context where the databasing is occurring, while also creating space and resourcing opportunities for ontic work to be continued by the Aboriginal experts off to the side, so to speak. In the collaborative ontic work, we are puzzling over what these digital objects being generated in the course of our cooperation *are*. And what they *do*. How strange they may become as we rethink them together. We do this work on the run, so to speak. Keeping the questions always explicit and open is crucial in dealing with fears that our easy modern assumptions about how to go on may compromise the Aboriginal knowledge work we are seeking to create space for.

Our other task is to decide what to do with the Aboriginal digital objects, how to arrange them, how to store them, how to find them and put them together. We need to agree on institutional arrangements, which inscribe the conventional ways we go about making and sharing knowledge. We need to do both tasks, and we need to do them separately. We need to remember how information architecture reflects, as much as it enacts, a politics of knowledge (Christie 2004, 2005a).

## How Do Epistemic Differences Arise When People Try to Work Technosciences and Aboriginal Knowledge Traditions Together, and How Is Postcolonial Databasing Involved?

One of the many reasons for researching the use of ICT in postcolonial databasing in particular and collective memory making in general is the need to identify and manage epistemic differences that emerge when knowers from differing knowledge traditions work together. An example of this is when environmental scientists and Aboriginal landowners try to work together to conserve biodiversity. In northern Australia perhaps the most valuable tool for ecological management is fire: "firing the bush." By maintaining a sophisticated regime of firings, Aboriginal land owners achieve a complex mosaic of dynamic ecological successions. Aborigines have been doing this in Australia for millennia. Science, which has been working with Australian nature for only a little over two hundred years, seems to be much less successful at achieving complex dynamic mosaics. Besides, much of the land in northern Australia is owned by Aboriginal Australians, who have a right to work their lands according to the standards of their own knowledge traditions.

A firing is judged to be a valid and efficacious instance of the knowledge tradition of Aboriginal Australians in several ways. These are epistemic concerns. Theories of knowledge determine the forms of witnessing and evaluating any instance of applying and engaging knowledge. In Aboriginal knowledge traditions, it is most important that particular knowledge authorities participate in specific roles in planning and executing the firing. Expressions of knowledge are not valid unless this condition is met.

The firing of any particular place must always begin in a carefully negotiated spot and proceed in certain ways, through a series of contiguous particular named spots in the landscape. The names of the series of contiguous spots in the landscape must be publicly and collectively recited before a firing begins. The knowledge authorities are those who know which spots are where, and the directions in which the various sequences of names move across the land. In addition it is important that particular items of food are gathered in the process of firing and distributed to appropriate persons in the correct relative amounts. This distribution of various foods collected from the multiple microecological zones that constitute the area fired (and adjacent areas, such as mangroves) expands the number of people who can attest a firing episode as legitimate. A particular firing will imply that people with particular rights are moving through places where it is recognized that particular foods are found. Being able to present the appropriate food items to others is a form of proof that the firing was valid. These institutionalized forms of proof and witness go along with an epistemology that sees that true knowledge can only be performed and enacted in place (Verran, 2002a, 2002b).

These forms of Aboriginal witnessing and evaluating an episode of firing differ markedly from the ways that epistemic concerns are institutionalized in environmental science, where justification is a separate moment from the collective work of firing. There scientists plan their firings with maps that allow areas to be delineated. They collect observations on the fire and its effects on vegetation and assemble the results in scientific papers that are published, reports that other environmental scientists might read. These reports attest and witness the efficacy of the firing. These forms express an epistemology that understands knowledge as representing an "out-there" reality (Law 2007).

Nowadays when it comes to managing Australia's northern savannas to promote a robust biodiversity through firing, there are two incommensurable standards. The Aboriginal way of judging a firing to be a valid expression of knowledge cannot have any salience from the point of view of Western ecological science. In the same way, it is inconceivable that scientific validation could be legitimate in Aboriginal traditions. The epistemic differences are irresolvable as such (see Verran 2002a, 2002b).

Yet it is possible to imagine that performative Aboriginal ways of doing their fire knowledge can be judged to be successful by the objectivist standards of Western science. This judgment has in fact been made many times and provides the imperative for Western fire ecologists who try to learn from Aboriginal "firestick farmers" in the first place. It is also possible (but more difficult) to imagine the fire abatement regime of a small local volunteer fire brigade of white Australian landholders to be judged occasionally satisfactory by Aboriginal elders. Many white Australians in the bush have learned to manage the fire issues on their land by imitating Aboriginal practices: burning early; remembering the fires from the last few years; knowing and loving every part of their land; learning to read the day-by-day seasonal changes in dew, afternoon winds, overnight temperatures, and so on; keeping in regular, honest contact with their neighbors; coordinating and helping with each other's firing.

Can digitizing technologies managed with a postcolonial sensibility help collaborations between Western and Aboriginal experts? It is possible to imagine assembling digital objects during the planning, execution, and evaluation of firing. Audio files can capture what is said, and still images, movies, and spoken commentaries might also be gathered, along with the foodstuffs and the measurements, as Aborigines and scientists go about their tasks of planning, executing, and witnessing firing episodes. Shared databases are place and time specific. Imagine storing these digital items in a structure-free digital matrix. In databasing terms, this implies that no distinction exists between data and metadata. We refuse that distinction for as long as we can, to keep open the possibilities of the epistemic being negotiated alongside the ontic. We can imagine Aboriginal and scientific experts working together, looking at satellite images of patches, judging the mosaics, reviewing the photos of the participants and the processes of decision making, talking about ancestral songs and academic articles. The scientists will surely be impressed by some of the Aboriginal digital artifacts and probably even want to configure some of them for their own evidential purposes (which, of course, they would be welcome to do if they could do so without compromising the ontological fluidity of the collection). And the Aboriginal experts will find ways to configure some of the collection's objects for particular performances, to carry on the work of keeping people and place together in mutual commitment and as further evidence for the truthfulness of the work they have done. And we may even hope that this collaborative work may slowly and slightly nudge the institutional arrangements outside the computer that until now have kept Aboriginal fire experts out of the difficult and important work of preserving the biodiversity of Australia's "top end."

## How Do Ontic Differences Arise When People Try to Work Technosciences and Aboriginal Knowledge Traditions Together in Postcolonial Knowledge Work and Databasing?

Different knowledge traditions make different assumptions about what there is. One set of clues we can get about this level of difference lies in the grammars and morphologies of different languages. Grammars and morphologies are deeply embedded in, and expressive of, the ontic. Another set of clues can be winnowed out by considering the everyday forms of generalizing we find working in a knowledge tradition. For example, we can look at what is involved in using numbers. In the case of Aboriginal Australian communities, we need to consider the generalizing that makes up their very different form of mathematics. Yet another set of clues can be found in the stories that peoples tell about the origins of the worlds they know and the things that fill those worlds. This is generally called the metaphysics of a knowledge tradition and is the subject of our next question.

Imagine a scientist watching an old Aboriginal man demonstrating the process of making fire by rubbing two sticks together. The old man has chosen sticks from bushes that look very different. He uses one as a base and cuts a notch in the middle. He uses the other stick like a drill bit. Seating it in the notch, he twirls the stick quickly between his palms. Gradually a pile of hot tinder accumulates, and when it is smoking, he tips this smoldering pellet into a nest of shredded bark, which, when blown on, breaks into flame.

Enthusiastic and interested, a scientist asks the names of the two bushes from which the fire-making sticks were plucked. It is clear to him that the plants are very different; they belong to different biological families. He is genuinely shocked when the old man insists that they are *really* the same. While the old man accepts that the plants might look different, he insists that what is important is that *logically* they are "the same one." The old man and the scientist have been confronted with an ontic difference.

Is this ontic difference resolvable? Yes, but only by opportunistically assuming the existence of a third translating domain. This move involves an ontology that is both and neither Aboriginal nor scientific. But this is not a meta-ontology. It is not an ontic domain that supervenes and contains the other two. On the contrary, it is an infra-ontology, an inside connection. It takes enough of what matters ontologically to Aborigines when they are dealing with firings, and enough of what matters to scientists when they are engaged in doing their prescribed burns, and institutionalizes the two, effecting among other things a separation of the ontic and the epistemic. Learning how to do this in on-the-ground situations is not easy because it involves working with contradictions in disciplined ways. It is particularly difficult for scientists, because contradiction is usually outlawed in science. The work in this infra-ontological space is essential empirical work centering on metaphysics. The key points here are that agreements about how to care for land by using fire can be reached by working the disparate traditions together, and digital resources can be useful for this work provided their categorization is not preempted by hidden Western ontological assumptions.

## How Does Postcolonial Databasing Emerge in Infra-ontological (and Epistemic) Work?

Let us look more closely at this infra-ontological work, which is necessarily also epistemic work, and consider how postcolonial databasing can contribute to its emergence and functioning. Sameness and difference are constituted through alternative framings in science and Aboriginal knowledge traditions. People are often shocked when they experience this metaphysical form of difference. Another similar sort of disconcerting experience is often associated with differences around the relation whole and part. Recognizing this sort of difference can also emerge when Aborigines and scientists try to learn each other's firing regimes.

Scientists assume that a thing like a habitat or ecosystem is a real entity found in nature. While the habitat may possess many attributes and characteristics that could be the subject of quite different scientific disciplines—pedology, botany, hydrology—the habitat itself is just a single given object. It has a particular out-thereness that lends it singularity (Law 2007, 600). Many different representations of this given, whole thing might be made, and they tell of the various aspects of a single thing. The differences between the experiences of the separate groups of scientists who research a habitat are downplayed and backgrounded. This being so, when scientists report their burning of an area, they tell about their activities in accordance with the assumption that they are about a single entity. They take great pains in the introductions and conclusions of their reports to show that all the separate experiences of the scientists were caused by the properties of one thing: the habitat under observation.

But when Aborigines report their episodes of burning, they completely (and deliberately and responsibly) fail to attend to the place as a whole—what they call *wäŋa* (land, place, home). They emphasize and recognize only the diverse involvements of the groups (of people, names, ceremonies, places, resources, totems, ancestral connections, etc.) who have various and variable interests at stake in a collective episode like a firing. The singularity achieved in different kin groups working together in a single, purposeful episode is the taken-for-granted background in any reporting. Aborigines do not assume that places exist in the here and now as single, whole things. Places might achieve a form of ephemeral singularity when a firing or some other such collective activity occurs—if all the correct people are present and things are done in a correct manner. Those ephemeral unities of actual existence are achieved reenactments of an originary act of creation by spiritual ancestors. For this reason, to organize our digital resources within a database under the category of "place," for example, might easily compromise the here-and-now ontic work that Aboriginal knowledge practices demand.

As scientists see things, reports of firings given by Aborigines are too fragmented and interested. By contrast, Aborigines feel that scientists fail to credit properly the multiplicities that inhere in (performed) place. This is another instance of ontic difference. It too, with care and caution, can be worked around well enough for Aborigines and scientists to feel confident in going on together so long as the ontics and the epistemics are kept separate. Viewed from the point of view of postcolonial databasing, ontologically speaking, the information architecture that will serve scientists' purposes in reproducing their traditions must be led to serve Aboriginal purposes, and vice versa. Difficult and counterintuitive though it might be, postcolonial databasing must then go to great lengths to avoid preemptive categorization—that is, databasing with prescribed metadata.

Scientists can (and do) use the work of talking with Aboriginal people about land (and associated digital objects) to validate their own work. Connections are always possible. But postcolonial databasing must always also allow for the possibility of total separation. Thus, in our experimental database, we began by planning for an ontologically fluid data structure (Srinivasan and Huang 2005). The only a priori distinctions we allowed were those between the different file types: texts, audio files, movies, and images (hence the name TAMI) (Christie 2005a, 2005b; Verran, Christie, Anbins-King, van Weeren, and Yunupingu 2007; Verran 2007).

## How Do the Sources of Aboriginal Knowledge Differ from Those of Technoscientific Traditions, and How Are the Differences Significant in the Context of Postcolonial Databasing?

The metaphysics of Aboriginal Australian knowledge traditions are very different from those of the technosciences. They have very different accounts of the origins of knowledge. In Aboriginal Australian traditions, knowledge is taken as already always in the land. However, knowledge needs the correct circumstances for true expression. In Aboriginal Australian knowing, there is no given or a priori separation of places and persons who belong to that place. Knowledge is in the land and in people by virtue of their belonging to the land (Verran and Christie 2007).

The source of Aboriginal knowledge/place/persons is often named in English as "the Dreaming." This is a transcendental space-time parallel to the secular time of the ordinary here and now. The sense in ordinary English usage that dreams are close by, a parallel of the everyday without the conventions of material space and time, is useful here, but the idea that dreams are somehow *not real* is definitely not part of the Aboriginal sense of the Dreaming. In Aboriginal metaphysics, the creative impulse for the world arose and continues to arise, its creative impulse emerging from the complex collective lives of a multiplicity of originary beings, both human-like and nonhuman in form. Entities that can be known in Aboriginal Australian knowledge are framed primarily as here-and-now expressions of the Dreaming. Knowledge and the spiritual life of religion are not separate in Aboriginal traditions, so all things have an explicitly recognized metaphysical dimension.

As well as an ultimate division between the eternal Dreaming and the secular here-and-now world of everyday individual experience, there is a subsidiary division between the world's two sides. Both the secular domain and the Dreaming are divided into formal opposites. Among the Yolngu Aboriginal clans in northeast Arnhem Land, for example, these two sides, or moieties, are named Yirritja and Dhuwa. Everything is either Dhuwa or Yirritja, which are ontologically prior to place and time. The Yirritja and Dhuwa originary beings walked across the land, bringing it into being. Knowledge in the ordinary world, of the secular, is the outcome of Dhuwa Dreaming knowledge and Yirritja Dreaming knowledge working together to generate true expressions of the Dreaming. Knowledge in the secular here and now (an episode of firing, for example) is justified as a true expression of the Dreaming if relevant knowledge authorities of the opposed moieties, with interests in the particular set of issues at hand, witness and attest a particular expression of the Dreaming as valid.

In the technosciences, while many practitioners might profess religious belief, Islamic, Buddhist, or Christian, for example, such metaphysical commitments are not embedded in the forms of justification of technoscientific knowledge. While the entities of technoscientific knowledge do indeed possess a profound and intrinsic transcendental element, in claiming to be free of such "taint," and in using particular configurations like space and time, this idea is hard to articulate, since knowledge of the world is taken as distinct from the world itself. Knowledge is made as representation. Yet it is important to acknowledge that the very gesture that separates world and its representations in Western knowledge traditions is itself transcendental in nature. Both Platonist argument and Abrahamic religion effect a distancing of human knowledge from divine creation, yet we moderns do not often remember this when we assume digital objects as representations.

The sciences make a fundamental division of people as *knowing* and things (including places) as *known about*. Things known are matter that extends in space and time and is situated in an empty space-time frame. In a secondary or derived way, abstract things like numbers are understood by analogy to primary material things. True knowledge about those material and abstract things is taken as accumulating through the application of proper scientific method. Knowledge is justified as true if it can be shown to have been produced in valid ways.

Despite the very different outcomes of the processes of articulating and justifying origins of knowledge in Aboriginal and Western traditions, transcendence manages the paradox inherent in both. Yet acknowledging the difference is crucial in postcolonial databasing. On the one hand, in Aboriginal Australian knowledge traditions, databases are implicated in effecting transcendence and recognizing the land itself as created by transcendent beings, as the source of knowledge. In Western technoscience traditions, databasing is caught up in the gesture that denies knowledge as transcendent and separates world and knowledge.

**Aboriginal Knowledge Is Located in the Land Itself. How Can Knowledge Be Stored in the Land and in Databases Too?**

How can we understand Aboriginal people when they say that knowledge is in the land? How can science learn how to take that claim seriously? And how is postcolonial databasing involved? The land is a set of sites with meaning embedded, with information there in place. But those meanings, necessarily organized in some way, are accessible only to those who have been sensitized and trained in the right traditions.

One way to think about databasing in an Aboriginal context is to understand a computer as a simplistic and "outside" version of one of those meaning-laden sites in land. In this way of understanding things, the database and the land are the same sort of object. Doing databasing can contribute to the remembering and forgetting that is inherent in community life, as can "doing ceremony," which mobilizes information embedded in the land. Databasing can be understood as a way of doing outside (as in public, nonsacred) collective memory with digitized materials. Images made with digital cameras (both video and still), audio files, written texts typed up on a computer, and so on, can record something that might be presented again later in another forum in such a way as to help those involved in some endeavor to remember (or learn). Seeing things in this way reminds us of the importance of developing some protocols around the generation of digital objects, as well as their storage in postcolonial databasing.

Particular significant forms of Aboriginal knowledge making center on ceremonies, some of which might, for example, involve episodes of firing the land. In much the same way, technoscientific knowledge making pivots around the workings of laboratories and field sites, which might also, for example, be involved with investigating firing of the land. Just as there are many and varied types of laboratories, so too are there many different sorts of ceremonies in Aboriginal life. These are religious ceremonies, but they are not repeated rituals. No two ceremonies are identical in Aboriginal life. Each is concerned with spiritual practice and knowledge making with respect to particular times and places and groups of people. It is their very difference, their special connectedness to the here and now (not the there and then), that forms the most fundamental criterion assessing the truth claims of the performers and choreographers. Knowledge management and resource management are mutually entwined (Christie 2007).

We can describe scientific knowledge making in laboratories and field sites by elaborating the specific sorts of social institutions involved, the material routines that are crucial in knowledge making, and the literary texts and literacies involved. But to give a complete picture, we need also to include the paradigms, theories, or imaginaries in which these processes make sense. These same headings can be used to describe the workings of Aboriginal ceremony and the knowledge making that occurs in them. Just as the entities that emerge from laboratories and field sites remake their worlds, so do the entities that emerge from the ceremonies of Aboriginal Australian life. In ritual and ceremony, Aboriginal knowledge authorities use many diverse sources of information. In ceremony, dance, painting, song, and story need to be performed correctly and under the right auspices to become knowledge making.

Often people see databases as archives, a sort of digital museum. But in our postcolonial databasing project, we are *not* treating databases as digital museums. We are asking if databasing can become a useful additional experience—an extra site of performance. Can digitized information feed into, complement, and extend the already well-developed ways that information is handled and managed in Aboriginal communities to support Aboriginal people in doing their knowledge? And would it be useful to think of databases as behaving thus in their modern uses? Under what conditions might databasing become a useful form of managing information understood in this way as flowing performance? These are empirical questions, and Aboriginal people are the ones who must drive the process to come up with answers in concert with practitioners of Western knowledge traditions. Recognizing what has already been achieved here, we point again to the issues associated with differentiating databases as representations and with understanding them as performances.

**Aborigines Have Local Knowledge. How Is Local Knowledge Consistent with Having Databases?**

The notion of databases as something other than local in application tends to derive from a vision of what we could call "the imperial archive" (Richards 1993). It conjures a picture of Jules Verne's *Nautilus*, tirelessly scooping up and organizing observations of the underwater world. This image assumes two things. First, it takes for granted the existence of "observed facts," or little pieces of knowledge referring to a single given reality; and second, it assumes that if you could only get enough of them together in one place, facts would eventually link up into one complete system of knowledge. This vision of databases as knowledge imagines them as inventories, audits, and collections. In many traditions of indigenous knowledge (and in many sciences), both assumptions are seen as the wrong way to understand knowledge.

Anyone who thinks about the notion of universality as just the nonlocal will see that observed facts are always generated and made solid in specific places and times by particular groups of people using particular means or knowledge practices. Knowledge is always done in specific ways. It is a commonplace that it is actually very difficult to get things to link up. It is sometimes very difficult actually to link working databases—for example, those that have been assembled in doing biodiversity (Bowker 2000). We also found this when we started searching for databases in northern Australia that included "indigenous knowledge." Any database *is* a form of local knowledge. It is a collection in digitized form of data items that have been generated using specific local methods. And this is all the more evident now that databases are scripting devices for managing Web applications—very local accumulations designed to be read by a Web browser.

Of course, Aboriginal knowledge is local. All knowledge is local. It remains true that sometimes, with prodigious collective effort, some or even many local knowledges can be linked. Sciences are often good at linking up their local knowledges, although sometimes it is difficult to get different sciences to work together. Sometimes and in some places scientific knowledge and Aboriginal knowledge can usefully be linked, as we elaborated earlier.

Nonetheless there *is* something to the claim that Aboriginal knowledge is distinctively local in that, being performative, it actively resists extension from particular time and place. We have dealt with this earlier in suggesting that postcolonial databasing is in part about learning to work with databases as coparticipants in local collective action.

## How Could Elements of Traditional Culture Be Strengthened by Supporting Aboriginal People in Their Use of Digitizing Technologies?

A problem arises if we think of traditional Aboriginal knowledge as antimodern, the opposite of modern culture. Then we will begin to think of traditional cultures as stuck in the past and want to put them in a museum and close the exhibit case. Understanding "traditional" in that way, we think of it as somehow inconsistent, perhaps even incompatible, with computers. Traditional cultures are contemporary forms of life just as modern cultures are. They are rich in modes of innovation, as well as having ways to preserve cultural forms. We can understand traditional cultures as involving nonmodern forms of identity. They have ontologies that make modern assumptions about knowledge and knowing look strange. Digitized information arranged in ways that make sense and are usable by those working within nonmodern cultures can surely be devised. As long as we do not make assumptions based on modern ways of using digital objects, if we proceed in open ways, collaboratively and empirically researching how indigenous people actually use digitizing technologies, we keep open the possibility of strengthening traditional forms of cultural innovation with computers (Christie 2008).

Traditional forms of passing knowledge from an older generation to a younger one often involve young and old being in the same place at the same time, doing things together, talking about them. They involve a process of reimagining going on together, finding new forms in which to express the understandings in sharing them. We often find that indigenous people want to assemble collections of digitized items for specific reasons. They want to be able to intervene in a specific context in a particular way. Assembling digitized items in these projects becomes a site, a time and place where young and old, with their varying competencies, work together. Databasing can become an impetus for young and old to work together in ways that can empower and educate the young while recognizing older people as knowledge authorities.

Protecting collective intellectual property is important in all "closed" knowledge economies. Aboriginal societies are no different from American corporations in this. The issue is controlling who knows and how much they know. Strategic revealing and hiding are involved. Modern companies protect their intellectual property with patent laws, by various technical means, and by selectively authorizing and commissioning various knowers. Aboriginal clans have equally effective means of managing the strategic revealing and hiding of intellectual resources.

Two separate elements need to be considered in thinking about intellectual property and indigenous knowledge with respect to collections of digitized items that point to natural and cultural resources. The first relates to forms of management for these collections that express indigenous ways of doing intellectual property. We need to find workable ways of respecting different clan ownership of various elements and recognizing differential individual access. Our stance at this point is to restrict our research to secular contexts. We avoid engaging with knowledge that is sacred and religious. Another workable strategy that emerged in our research was the insistence of our coresearchers that they keep their own personal databases and negotiate sharing for each particular new knowledge production exercise that arises—on a case-by-case basis. Second, maintaining collections of digitized material in ways that protect the collections appropriately to avoid piracy from outside interests is important. The risks of piracy were mitigated by the Aboriginal insistence that their objects not be stored on the Internet—not even in the databases of the community knowledge centers. We were able to resist the pressure from programmers for the "future proofing" "interoperability" and "extensibility" of the system we developed (Christie 2005a).

## Can We Articulate Some General Principles for Postcolonial Databasing and More Generally for Thinking about Engaging Disparate Knowledge Traditions through the Mediation of ICT?

Genuine recognition of difference can be painful. It involves beginning to doubt our own knowledge traditions as sources of absolute certainty and see them as having limits. Accepting that every knowledge tradition is inherently and systematically partial is challenging. It is sometimes difficult to accept the profound significance of difference and at the same time persevere in learning about the other and in considering how our familiar ways of knowing might engage with other ways of knowing. Very often we approach other knowledge traditions thinking that they are just an odd or unusual version of the ways we know. That is a form of inauthenticity.

The odd aspect of seriously engaging with the other is that to recognize difference in knowledge traditions, we need to "make strange" our own. In part that is what we have tried to do in these questions and answers, telling of some of the issues that arise when Aborigines and scientists work together and mediate that working together with digital objects. Beginning to explore what postcolonial databasing might be, we engaged in a process of "strangification." We made strange the epistemological assumptions of science, revealing them by setting them alongside another way of witnessing valid expressions of knowledge associated with an alternative account of truth. To make strange our own knowledge traditions, we must begin to open up questions of metaphysics.

Postcolonial databasing is doing a form of experimental metaphysics. This is to make each side strange with respect to itself so as to find ways to connect with the other. An experimental metaphysics is a framing of issues of difference that takes elements of both metaphysical systems to develop what we might call an ad hoc hybrid translation borderland. It can provide a way to imagine how we might connect in partial, strategic, and opportunistic ways. Some entities that might usefully be linked in partial ways—like Aboriginal firings and the prescribed burns of science—can be identified as simultaneously distinct and connected. The on-the-ground activities that enable strategic linking can be identified. Each firing can begin to make some sense in the other knowledge tradition through the use of a metaphysically explicit translating zone. This can help us begin to accept the limits of our own ways of being certain about what we know—our own types of epistemic standards. And postcolonial databasing can be a means for promoting such ontic and epistemic fluidity.

## References

Bowker, G. 2000. Biodiversity datadiversity. *Social Studies of Science* 30 (5): 643–683.

Bowker, G. 2006. *Memory Practices in the Sciences*. Cambridge, MA: MIT Press.

Christie, M. 2004. Computer databases and Aboriginal knowledge. *Learning Communities: International Journal of Learning in Social Contexts* 1:4–12.

Christie, M. 2005a. Words, ontologies, and Aboriginal databases. *Multimedia International Australia* 116:52–63.

Christie, M. 2005b. Aboriginal knowledge traditions in digital environments. *Australian Journal of Indigenous Education* 34:61–66.

Christie, M. 2007. Indigenous knowledge management and natural resource management. In *Investing in Indigenous Natural Resource Management*, ed. M. K. Luckert, B. M. Campbell, J. T. Gorman, and S. T. Garnett. Darwin: Charles Darwin University Press.

Christie, M. 2008. Digital tools and the management of Australian desert Aboriginal knowledge. In *Global Indigenous Media: Cultures, Practices, and Politics*, ed. P. Wilson and M. Stewart. Durham: Duke University Press.

Hamilton, Carolyn, Verne Harris, Michele Pickover, Graeme Reid, Razia Saleh, and Jane Taylor. 2002. *Refiguring the Archive*. Dordrecht: Kluwer Academic.

Law, J. 2007. Making a mess with method. In *The Sage Handbook of Social Science Methodology*, ed. W. Outhwaite and S. Turner. London: Sage.

Richards, Thomas. 1993. *The Imperial Archive: Knowledge and the Fantasy of Empire*. London: Verso.

Scott, Gary. 2004. Audit of indigenous knowledge databases in Northern Australia. http://www.cdu.edu.au/centres/ik/pdf/IEK_Audit_Report24_06_04.pdf.

Stengers, Isabel. 2005. The cosmological proposal. In *Making Things Public: Atmospheres of Democracy*, ed. Bruno Latour and Peter Weibel. Cambridge, MA: MIT Press; Karlsruhe: ZKM/Center for Art and Media in Karlsruhe.

Srinivasan, R., and J. Huang. 2005. Fluid ontologies for digital museums. Special issue, *Journal of Digital Libraries* 5 (3).

Suchman, L. A. 2002. Located accountabilities in technology production. *Scandinavian Journal of Information Systems* 14 (2).

Verran, H. 1998. Re-imagining land ownership in Australia. *Postcolonial Studies* 1 (2): 237–254.

Verran, H. 2002a. A postcolonial moment in science studies: Alternative firing regimes of environmental scientists and Aboriginal landowners. *Social Studies of Science* 32 (5–6):1–34.

Verran, H. 2002b. "Transferring" strategies of land management: The knowledge practices of indigenous landowners and environmental scientists. *Research in Science and Technology Studies* 13:155–181.

Verran, H. 2007. Software for helping Aboriginal children learn about place: The educational value of explicit non-coherence. In *Education and Technology: Critical Perspectives and Possible Futures*, ed. David W. Kritt and Lucien T. Winegar, 101–124. Lanham, MD: Lexington Books.

Verran, H., and M. Christie. 2007. Using/designing digital technologies of representation in Aboriginal Australian knowledge practices. *Human Technology* 3 (2): 214–217.

Verran, H., M. Christie, B. Anbins-King, T. van Weeren, and W. Yunupingu. 2007. Designing digital knowledge management tools with Aboriginal Australians. *Digital Creativity* 18 (3):129–142.

# 5 Sacred Books in a Digital Age: A Cross-Cultural Look from the Heart of Asia to South America

Hildegard Diemberger and Stephen Hugh-Jones

## Introduction

The use of digital technologies in the reproduction of texts has profoundly transformed attitudes toward books and also the production and format of books themselves. Remarkably, at a time when it is often suggested that digital books may soon supersede conventional books, there has been a lot of rethinking about what a book is, as both object and artifact (Boutcher 2012, 59). In addition it is often assumed that digital technologies will lead inexorably to a universalization and standardization that will override cultural differences and local identities. The way in which digital technologies are used in relation to books appears, in fact, to be predicated on what a book represents in a particular context. By looking at a range of ethnographic cases from Asia and South America, this paper examines culture-specific understandings of books, book-related technological innovations, and digital objects. It will show that we cannot take for granted that a book is simply a conveyer of a message; books may have many more dimensions that raise interesting questions: people do things to books, but what do books do to people, and can a digital derivative do the same? Did the Buddhist understanding of texts as relics to be propagated, which inspired the invention of printing in the seventh century (Barrett 2008), have an impact on the current adoption of digital technologies in Buddhist societies? Is a digital derivative of a sacred book also sacred? What is the role of books and digital objects among people who were, until recently, "peoples without writing"? Do cultural factors predispose some peoples to favor the adoption of books and others to favor alternative technologies such as VCRs?

As are the texts of many other literate cultures, Tibetan and Mongolian books are being rediscovered, cataloged, and scanned in libraries, museums, and monasteries around the world. We begin by outlining the Tibetan-Mongolian Rare Books and Manuscripts (TMRBM) Project funded by the Arts and Humanities Research Council (AHRC) and British Library and hosted by the Mongolia and Inner Asia Studies Unit (MIASU) at the University of Cambridge. This project aimed at enhancing the preservation, availability, and understanding of some 2,500 Tibetan and Mongolian books and manuscripts held in British libraries, including the entire Younghusband collection, some 1,500 volumes gathered by the British scholar-soldier Francis Younghusband during the 1903–04 invasion of Tibet. This project was also the starting point for spin-off projects that linked up several similar enterprises in Tibet, Mongolia, and Bhutan and led to a deeper exploration of the nature of books and technological innovation in the Buddhist context. With the creation of large, interconnected databases, the accessibility of texts has acquired an unprecedented dimension, enabling people to reconstitute collections and volumes that have been broken up and dispersed; it has also become part of the current revival of Tibetan Buddhism and the restoration of the Tibetan cultural heritage after political disruption and religious persecution. By looking at the ambitious vision underpinning these enterprises, as well as their practical problems, this paper also raises some of the dilemmas of digitization as a transnational medium for archiving and dissemination. Finally, following the experiences of Stephen Hugh-Jones, the head of the TMRBM Project and one of the authors of this paper, we will outline how an Amazonianist, becoming involved in this book-centered undertaking, was prompted to rethink the role of books and the uses of digital technologies among peoples without literate traditions. If digital technologies are part of an international process of rediscovering, cataloging, and preserving Tibetan and Mongolian books and manuscripts, in northwest Amazonia these same technologies underwrite an unprecedented surge in publishing. Here, instead of acting as mere "informants" for works published by anthropologists, indigenous authors have now begun to publish their own traditions in their own names.

On the basis of diverse experiences, we thus show that the understanding of a literary artifact, and of how it may relate to an oral tradition, shapes the modes in which it relates to digital technologies. This may hold the secret of why the life of conventional books may still be long and varied, alongside their multifarious digital incarnations that make texts of different traditions accessible across the globe in an unprecedented way.

**The Tibetan-Mongolian Rare Books and Manuscripts Project: Anticipated and Unanticipated Consequences**

When the Dalai Lama went into exile in 1959, after a large-scale Tibetan uprising was crushed by the Chinese government, he was followed by some one hundred thousand Tibetans. They brought along what they considered most precious, and many took with them their sacred books. After the exile community had established itself in India, an intense reprinting activity of ancient Tibetan texts started. Gene Smith, who was then working in India for the Library of Congress, played an important part in this process and became a legendary figure in Tibetology. He eventually established the Tibet Buddhist Resource Center (TBRC), which makes an increasing wealth of Tibetan resources available online, and which he directed until his death in 2010.

Zenkar Rinpoche is an eastern Tibetan lama, and after the Cultural Revolution, he took a keen interest in the recovery and restoration of the Buddhist resources that have survived the turmoil in the Tibetan and Mongolian areas of China. He therefore linked up with other Tibetan scholars who shared the same vision, as well as with international scholars such as Gene Smith, and together they created an expanding global network centered on recovering and making accessible Tibetan resources worldwide. Books and fragments of books that had been scattered through colonial history and the Cultural Revolution could thus be reconnected as if they were part of a global jigsaw puzzle. Their very existence had the power to mobilize people both locally and globally, exercising an irresistible moral pressure, both in religious and secular cultural terms, and giving rise to diverse networks—including the Tibetan-Mongolian Rare Books and Manuscripts Project.

In April 2003, in New York, Gene Smith and Zenkar Rinpoche mentioned the Tibetan materials kept at the Cambridge University Library to the anthropologist Hildegard Diemberger, who had recently moved to the University of Cambridge. They asked about the Younghusband collection, which Gene had had the chance to browse in the 1960s. He knew that the materials collected in Tibet by Dr. Austin Waddell, medical officer and archaeologist with the 1903–1904 British mission, had then been split among different institutions, Cambridge University Library, the Bodleian at Oxford, the British Museum, and the India Office in London. He mentioned several important works that had been torn apart by this double process of partition, recalled some unique texts from this collection that had already been the object of research, and then mentioned other items that had so far not been captured by the radar of Tibetologists, such as several early block prints from the ancient Gungthang kingdom. They suggested that it would be wonderful if these materials could be reassembled, using both traditional and new technologies, to make them more readily available to Tibetans across the world and to the international community of scholars. The presence of these books, lying dormant in the library, while so much was going on in retrieving Tibetan resources, meant that concrete action needed to be taken. Hildegard, as an anthropologist, was at first overwhelmed not only by the scale of such a project but also because such a Tibetological endeavor was not an obvious part of her research and teaching focus. The awareness of the neglected materials proved, however, to be so compelling that when she came back from New York to Cambridge, she mentioned the Tibetan books to her fellow anthropologist Stephen Hugh-Jones, an Amazonianist who at that time happened to be head of the management committee of the Mongolia and Inner Asia Studies Unit (MIASU), itself part of the Department of Social Anthropology.

Initially Stephen's associations with MIASU and Tibet were largely contingent: having been head of the anthropology department meant a link with MIASU, and a previous visit to Bhutan had inspired a long-term but amateur interest in the ecology and cultures of the Himalayan region. However, as will be explained hereafter, what was initially simply a formal and administrative involvement with the TMRBM Project turned out to be directly relevant to his research among the Tukanoan-speaking Indians of northwest Amazonia.

Gene and Zenkar Rinpoche's suggestion was readily taken on at Cambridge, and a few weeks later, they were invited there to make a preliminary assessment to prepare an application to the AHRC for microfilming, digitizing, and cataloging the entire collection. The hub for the project was going to be MIASU at the University of Cambridge in cooperation with the institutions in Cambridge, Oxford, and London that housed the Younghusband collection. The team was gradually enriched with the curators of the British Library and the Cambridge University Library, the chairman and the directors of the management committee of MIASU, other consultants from Tibet and Inner Mongolia, and an energetic administrator. Karma Phuntsho, a Buddhist scholar from Bhutan, would be the center of the project, taking charge of the actual day-to-day work of research, cataloging, and supervising the photographic units at the different sites. The scope of the project also widened beyond the original idea of virtually reassembling the Younghusband collection, for the project was also in a position to act as a pilot for a UK union catalog of Tibetan resources and cover a parallel work on Mongolian materials.[1]

More than one hundred years after its arrival in the United Kingdom, the Younghusband collection had now started to bring together a heterogeneous and transnational team of people, all sharing the idea of making it globally accessible and enabling a virtual return of the texts, at least the most important ones, to their land of origin. These books, which had survived the upheavals of Tibet's recent troubled history thanks to a controversial operation of colonial collecting, could eventually participate in the current movement of Buddhist revival and preservation of Tibetan culture.

When the British mission returned from Tibet, after signing a nominal treaty, Waddell brought back with him "300 mule loads" of about two thousand volumes of religious books. The largest number of the books were sorted, listed, packed, and later shipped to the United Kingdom, where the collection was distributed between the libraries of leading UK institutions. These books and other Tibetan books at Cambridge University Library, the World Museum in Liverpool, and the Ancient India and Iran Trust form the scope of the Tibetan-Mongolian Rare Books and Manuscripts Project. After microfilming and cataloging, the selectively digitized resources have gradually been deposited in D-Space at Cambridge—an MIT-supported digital object management scheme—with the expectation that this repository will guarantee their long-term preservation and accessibility.

Beyond the planned work on the collection, this initial project turned out to be the starting point of a chain of projects, which increasingly linked what was going on at Cambridge with preservation and digitization projects that had been developing in the Tibetan areas of China, India, and Bhutan. Encountering monks and lay scholars involved in such projects was a revelation for most members of the Cambridge team. They became progressively more aware of numerous formal and informal, secular and religious, local and transnational, networks centered on recovering, cataloging, and reprinting books and driven by the moral pressure that the very existence of these surviving books could exercise on them. They also realized that if members of these networks had readily adopted all that new technological developments could offer to facilitate and enhance their task of dealing with their precious texts, it was also the case that the way in which these technologies were used was informed by the particular cultural significance of the material in the first place.

In this way, the initial project, which brought to light some unique Tibetan early prints, also instigated further reflection on the nature of Buddhist books and on the transformative process that was unfolding with the introduction of digital technologies in the handling of textual collections. What had started as an enterprise largely concerned with Tibetology and librarianship was gradually acquiring an anthropological dimension that eventually took the shape of a further AHRC project titled "Transforming Technologies and Buddhist Book Cultures: The Introduction of Printing and Digital Text Reproduction in Tibetan Societies."[2]

## Transforming Technologies and Buddhist Book Cultures: What Is a Book in the Context of Tibetan Buddhism?

In both the Mongolian and Tibetan languages, holy books are addressed as honored persons—not as mere objects—and by their sheer presence often prompt people to find, restore, and look after them, thereby fulfilling a moral obligation. The moral obligations entailed in Buddhist books are best illustrated by verses reported at the end of *The Perfection of Wisdom in Eight Thousand Verses*: "When, through the Tathāgata's sustaining power, it has been well written, in very distinct letters, in a great book, one should honour, revere, adore and worship it, with flowers, incense, scents, wreaths, unguents, aromatic powders, strips of cloth, parasols, banners, bells, flags and with rows of lamps all round, and with manifold kinds of worship" (Conze [1973] 1994, 299). On the basis of these observations, we thought that, by referring to anthropological theories of nonhuman agency, it might be possible to gain a better understanding of what books are doing to people and how they can mediate a range of social agencies. We thus referred to Alfred Gell's theory that tried to capture the agency of artworks in a cross-cultural perspective. According to Gell, artworks and other inanimate objects such as stones or plants or body parts used as relics can be agents in an indirect sense, for although they themselves are not intentional beings, they can act as the medium through which people as human agents manifest and realize their intentions. They can become "extensions" of persons as part of their "distributed personhood."[3] Gell takes a broad view of what should be included under the label of "art object," but he does not include books among the items he discusses—presumably because books are by nature heavily verbal, a quality that might seem to run counter to the explicitly antilinguistic tone of Gell's argument. However, as Boutcher (2012, 59) has noted, in principle there is no reason why books should not be considered as art objects in Gell's terms.

It seems to us that Gell's approach can fruitfully be applied in relation to Tibetan books as an incarnation of Buddha's legacy. Since the dawn of Buddhism in India, books have been the repositories of Buddha's word. They embodied the teacher after he had passed away, a point that is stated already in the early scriptures such as the *Mahāparinibbannasutta*: "Then the Bhagvan addressed the Venerable Ānanda: 'It may be, Ānanda, that some of you will think, 'The word of the Teacher is a thing of the past; we have no Teacher.' The Doctrine and the Discipline, Ānanda, which I have taught and enjoined upon you is to be your teacher when I am gone" (Clarke Warren 1984, 107).

The same idea is reflected in *The Perfection of Wisdom in Eight Thousand Verses*,[4] which further emphasizes the merits accumulated by reproducing and distributing the text, as well as in many statements that Hildegard came across in contemporary Tibetan monasteries. Books are addressed with a language that is the same as the one used for icons and relics, and they are handled in similar fashion in ritual practice. In fact, icons, books, and relics or reliquaries are respectively symbols or receptacles (Tib. *rten*) of "body" (Tib. *sku*), "speech" (Tib. *gsung*), and "mind" (Tib. *thugs*) of the Buddha or of a later personality who continued the deeds of the Buddha. The idea of scriptures as "receptacles of speech" captures the fact that, read and recited by a human actor, the written text creates the audible presence of a person across space and time, just as effigies evoke the visual impact of the body and relics the spiritual power of the mind.

This view goes back to Buddhist India, where scriptures were often treated as equivalent to relics and, like relics, were used to distribute Buddha's legacy as widely as possible. They were all "expressions and extensions of the Buddha's biographical process" (Strong 2004, 5–10). The idea that a text could function like a relic, as a form of Buddha's presence, was later followed and further elaborated in China and throughout East Asia (Barrett 2008, 66). This was an essential element in the cosmopolitan cultural background of seventh-century China under the Tang dynasty, against which, according to Tim Barrett, the discovery of printing took place thanks to a Sogdian Buddhist monk, Fazang (643–712 CE), and the famous empress Wu Zetian (625–705 CE). Barrett observes:

> The notion of Buddha's words as relics was of particular importance to China's rulers, who needed to calm the fears of their subjects by demonstrating that they could distribute relics in large numbers. An empress of the Tang dynasty ... was particularly interested in ways of creating a large number of relics. Towards the end of her life in 705 she would also have been interested in multiplying copies of a "great spell," a newly translated text promising to its propagator extraordinary benefits in this life and the one to come. Since it is this work that turns up by the mid-eighth century in printed form in Korea and was printed in millions of copies by another empress in Japan, the presumption must be the text was actually distributed across East Asia in quantity by printing either in 705 or more likely as part of funerary rituals for the former empress in 706. (Barrett 2008, 131–132)

According to Barrett, the understanding of text as relic to be multiplied and distributed in a merit-making ritual had a significant impact on the discovery of printing. Centuries later this attitude promoted the adoption of printing technologies in Tibet, where the word for "print" (Tib. *par*) is the same as the word for "image." This attitude still informs many aspects of

Tibetan ritual life, from the cult of relics to prayer wheels and most prominently the *lungta* (*rlung rta*), the multicolored squares of cloth printed with an image of a divine horse and Buddhist formulas. Waving in the wind on top of houses and high passes, these have become not only the ubiquitous element of a religious custom but often also an indicator of Tibetanness. Commenting on their worship of the written word, Jack Goody (1968, 16) defined Tibetans as grapholatrists.

If books and other texts are receptacles of the Buddha's words, these words can be activated, released, and made to speak by various means—not just by reading with the eyes but also by physically acting on, or moving in relation to, such receptacles. Prayer beads, prayer flags, and various kinds of prayer wheels powered by wind, water, human hands, the hot air from burning butter lamps, or miniature solar panels are all examples of text objects that are made to be set in motion. The effects of this motion are further accentuated by various kinds of replication internal to the object itself. Beads and flags are ranged in repetitive series along a string, line, or pole, and the whirling drums of prayer wheels not only have a mantra written or embossed on the outside but also contain within them many meters of text on a scroll tightly rolled around a spindle that is itself inscribed with yet more words. Alternatively whole libraries of books may be arranged in the revolving bookcases found in Tibet and other parts of China.[5] Finally motion can be supplied not by moving the text object itself but by moving in relation to it. In their circumambulations, pilgrims carrying prayer wheels and prayer beads walk clockwise around temples that are repositories of books, around *mani* walls made up of stones inscribed with mantras, and around books themselves. Many of these pilgrims are illiterate, but through their motions, they can still activate the written word.

In each case, the principle is one of reduplication and repetition: as many words, as many lines, as many revolutions, and as many repetitions of the same text as possible. It follows from this that, potentially, CDs, videodiscs, and computer hard drives could all belong in the same class of rotating text objects, but here with the additional advantage of yet more bytes, yet more words, yet more text and even higher speeds, a point that has also crossed the minds of many in the Buddhist world. A Buddhist website, dharmahaven.org, tells us: "His Holiness, the Dalai Lama, has said that having the mantra on your computer works the same as a traditional Mani wheel. As the digital image spins around on your hard drive, it sends the peaceful prayer of compassion to all directions and purifies the area." The same website offers information on how to install several types of digital prayer wheels on your computer:

Digital videos of prayer wheels
Turn your hard disk into prayer wheels
Animated prayer wheels for Web pages
Download a prayer-wheel screen saver

Consistent with these links between speaking words and moving bodies, books are also treated as persons. They are clothed in robes (Tib. *na bza'*) like Buddhist monks and receive the prostrations of pilgrims who touch them with their heads to receive blessings. In springtime books are also paraded through the valleys and invited to bless the land and crops. As repositories of the sacred Dharma, the words of the Buddha and other great teachers, books are identified with their persons, as much relics as reading material. While distributing and multiplying has been crucial for certain Buddhist texts, some esoteric teachings, considered to be secret, are passed on only from teacher to disciple in a well-established relationship. In Tibetan societies in which patrilineal clans and lineages play a role, such as the Sherpa, books of secret teaching can be passed on from father to son, following "bone lines."[6] In this case, their handing down from one generation to the next follows forms of social organization rooted in Tibet's pre-Buddhist notions of kinship and bodily substances, and books may be dealt with as ritual objects that link up with ancestral cults.

One of the effects of printing in the West has been an increasing separation between body and text, speaking and reading, and a disjunction between writing as an integrated manual and intellectual craft and writing as abstract composition divorced from the manual work of typesetting or keyboard inputting (see Ingold 1968). This separation was always less evident in the Tibetan Buddhist world, where much of writing was concerned with recording the words of the Buddha and later commentators, where words are chanted as they are read, where the manual copying of texts was part of the training of many monks, and where the ongoing craft of woodblock carving produces a script that is close to the written word. The act of printing from a block not only reproduces a text but also transfers a blessing from the matrix to the printed page just as relics such as teeth or effigies can be used to mark dough or clay objects that are thus imbued with the blessing. Movable type, which was introduced only late in Tibet and to a very limited extent, disconnects these interconnections, and in fact a book printed with movable type is generally considered less sacred than a handwritten or block-printed one, even when the message or content of the book is exactly the same.

Digitization opens up a new space in which to negotiate the significance of objects, for it both severs some of the traditional links between person

**Figure 5.1**
Monks with computer. Photo by Hildegard Diemberger.

and text but also enables a reproduction that is conceptually closer to block printing than movable type. This is dramatically illustrated in the institutions that have sprung up in Tibet and other parts of China to preserve and digitize Buddhist scriptures. In these places, as if to undo the potential alienation from a foreign technology and reappropriate the meaning of the technological process, trainee monks and nuns sitting in front of serried ranks of computer screens rock back and forth, chanting the words of ancient texts as they input these same texts and transform them into digital format. Next to them, young boys and girls, usually dropouts from the school system, learn how to read ancient Tibetan texts and type them into the computer. Although officially a purely secular cultural endeavor recognized and authorized as such, the whole operation is led by Buddhist scholars funded by private sponsors, usually wealthy Tibetan traders who reenact an ancient form of Buddhist patronage by funding the reproduction of texts as merit-making activity. The output has different formats, ranging from the loose leaves of the *pothi*-format volumes that appeal more to the traditional Buddhist audiences and are better suited to ritual use, to codex-shaped books that are less sacred and more apt to modern forms of scholarship, to digital objects. Digital objects have even started to acquire ritual efficacy: I have seen CDs wrapped in book "robes" and placed on the altar, even if I have never seen them carried around the fields to bless them. More generally, digital technologies, having become central to global efforts to preserve and distribute Buddhist scriptures, are now inherently part of Buddhism's ritual world of beliefs and practices. Something that is clearly expressed in a few verses of praise by Chokyi Nyima Rinpoche, reported in the recently published brochure *Tibetan Buddhist Resource Center: Creating a Digital Path to Tibetan Literature*:

I'll be doing prostrations every morning to this computer.
Thank you so much.
You are giving all of us a huge gem, a jewel and a gem.

Certainly the way in which digital technologies have been adopted in Buddhist contexts is deeply influenced by attitudes toward the means of communication that preceded them, while raising a whole range of new issues.

## Permanence and Impermanence: The Dilemmas

Buddhist doctrine emphasizes the impermanence of all worldly things. Logically this impermanence also applies to the books and manuscripts that incarnate the teachings of Buddha and his followers. However, manuscripts and printed books are one way of ensuring accurate and uninterrupted transmission of the doctrine. Paradoxically, the custodians of this doctrine, many of them monk-librarians, must also be concerned with issues of permanence as they seek to ensure the preservation of texts. Texts written on stone, on palm leaves, on birch bark, and on paper made of rags or plants such as Daphne or Stellara have all proved to be remarkably durable; the earliest Buddhist texts in the British Library date back to the first century CE. The advantages, in terms of speed, accessibility, and dissemination, of new media such as microfilm and digitization have to be set against their capacity to survive across the centuries, a capacity that is as yet unknown and untested. Here digitization presents special problems due to its reliance on electrical power and it being subject to rapid changes in format and technology. If the floppy disks of the 1980s are no longer readable today, will our new digital archives still be readable fifty years from now? Rather than replacing one technology with another, the best solution appears to be to mix media, to exploit the advantages of each, and to be aware of their respective shortcomings. Today publishers and libraries increasingly use digital media, while Buddhism, Hinduism, Islam, and other major religions use the Internet to communicate with bodies of the faithful scattered across the globe. Monastery libraries in Tibet are also using digital technology to catalog their holdings and to reconstitute, in virtual form, whole runs of precious works that were fragmented in times of political upheaval. Like them, the TMRBM Project has extensively used microfilms, XML databases, catalogs, digital scans, and the Internet. Indeed, it is a product of such media.

Digital media have many advantages over the older materials and techniques used to store, catalog, and disseminate the written word. They are smaller and more compact, more easily stored and searched, and can be used to disseminate copies of texts to a much wider audience, faster, more cheaply, and without damage to the originals. But compared to more traditional media, the familiar printed books in Western libraries or the woodblock-printed books wrapped in textile coverings that are characteristic of the Tibetan Buddhist world, these new digital alternatives to old techniques also raise some burning questions. Can they resist unauthorized tampering? Will they last? What happens when the intimate links between body, gesture, voice, and text are broken and replaced by keyboard and flickering screen? And what happens to old power structures and claims to special esoteric knowledge when everyone has free access to written materials and can read what he or she likes?

**From Tibet to Amazonia**

As mentioned earlier, Stephen's involvement in the TMRBM Project, and in the further research on books, printing, and digital technologies to which the project has led, turned out to have direct relevance to his own research among the Barasana and other Tukanoan-speaking peoples of northwest Amazonia. Initially one might suppose that Tibet and Amazonia represent two extremes so opposed in terms of geography, history, and culture that they could share little or nothing in common. Apart from everything else that divides them, the Buddhism that is characteristic of Tibetan peoples is intimately linked with books and literacy, while the Amazonian Indians are, by tradition, "peoples without writing." Why, then, might research in the one area be relevant to the other?

The first link lies in asking questions about books as objects in themselves in addition to being interested in their contents. We have already suggested that, in the Tibetan world, books have special qualities that are intimately linked with Buddhist culture and religion; books have aspects of personhood and function as ritual objects and as relics. It is, in part, this that explains the high status accorded to books in a population the majority of whom were, till recently, largely illiterate. We have also suggested that the special qualities attached to books and printing in the Tibetan world today also inflect Tibetan attitudes to digital technologies.

The past twenty years have seen a remarkable boom in book publishing among the peoples of the Upper Rio Negro (URN) in Brazil, with more than ten books all published in the names of indigenous authors.[7] The bulk of these works are by Tukanoan-speaking groups from the Uaupés River and have been published in the Coleção Narradores Indígenas do Alto Rio Negro (NIRN), a series produced under the joint sponsorship of the Federação das Organizações Indígenas do Alto Rio Negro (FOIRN) and the Instituto Socioambiental (ISA), a Brazilian NGO dedicated to defending the environmental rights and cultural heritage of indigenous peoples. Written in Portuguese, the books mostly provide an account of the origin stories, mythology, and recent history of a particular Tukanoan-speaking group, usually from the point of view of one particular clan. They are published in the name of the clan and group and also in the names of two indigenous authors, a knowledgeable *kumu* (priest-shaman) acting as principal source and his son, a literate teacher or member of a local indigenous organization acting as translator and scribe. In addition, an anthropologist has been involved in each book, helping with the recording, editing, and publication and supplying an introduction and explanatory footnotes.

As this boom in indigenous publishing has no equivalents elsewhere in Amazonia, and as significant cultural differences exist between URN peoples and other Amazonian Indians, this raises some interesting questions: why should URN peoples be so interested in publishing, and is there something about books in particular that makes these new objects fit into existing cultural expectations? It appears that, for the Tukanoans, books can be seen as transformations of traditional ritual objects that can also be considered as relics, and, among other things, publishing can be seen as a transformation of previous ritual activity.

The second link lies in the way that digital technologies can be used to counter the fragmentation and loss of cultural heritage that results from political disruption and religious persecution, a process that is linked with reassertions of ethnic identity. As explained earlier, in the Tibetan world, digital technologies are being used as part of an effort to undo some of the effects of the Cultural Revolution. The Tukanoans of the URN also suffered their own form of cultural revolution, this one led not by political extremists but by Catholic Salesian missionaries who, over the course of the last century, carried out a program of religious persecution focused on the older generations and taught the younger generations to read and write and to

**Figure 5.2**
Tariana book. Photo by Instituto Socioambiental.

be ashamed of the culture and traditions of their elders. The Tukanoans' interest in publishing can be seen as an outcome of religious persecution in a specific cultural context in which books have a special resonance both as fitting in with existing cultural patterns and as part of a more general effort to reverse a perceived loss of culture and to reassert indigenous identity. To pursue these arguments, we must first provide some brief details of Tukanoan culture.

## Books as Ritual Objects

Tukanoan culture and society are based on an interplay between principles of similarity and difference. On the one hand, all URN Tukanoans see themselves as the same kind of people who share a common origin and a distinct culture and set of values. This common identity is underwritten by a shared body of sacred origin stories and other narratives. On the other hand, the Tukanoans are divided into some twenty separate exogamous, patrilineal groups, each speaking its own language and each defined by a common patrimony passed down the male line. In part, this patrimony or ceremonial wealth (*gaheuni*)[8] consists of flutes and trumpets, feather ornaments, stools, gourds, cigar holders, and other items of sacred ritual property. In the narratives of creation and origin that are now being published in books, these objects appear both as ancestors and as the body parts of these ancestors—musical instruments as bones, stools as pelvis, cigars in holders as legs, pelvis, and penis, and so on.[9] In this sense, these lineally transmitted objects are heirlooms and relics, features that bring them into line with Tibetan books.

The material patrimony of each group is complemented by different forms of immaterial verbal property: the language its members speak plus a set of personal names, songs, spells, and ritual chants, all of which are encompassed by particular versions of the origin narratives mentioned earlier. These origin narratives are at once the genealogy of each group and the pedigrees of their sacred material property, and they serve to affirm territorial rights, claims to status, and various ritual prerogatives associated with the group in question. This means that the narratives are inherently political statements.

Three points should be noted here. The first is that the hierarchical emphasis and patrilineal ideology common to the peoples of northwest Amazonia set them apart from the egalitarian and cognatic pattern of social organization typical of most other indigenous Amazonians. The second point is that this ideology is not one of blood but one of words and things:

the inheritance of a common material and verbal patrimony, not a line of descent or blood, defines the groups concerned. Today Tukanoans identify this common patrimony of ritual objects and sacred narratives with Western notions of "culture" and speak of their different groups as marked off by differences in "culture." In this sense, it might be said that Tukanoans are preadapted to the idea of "culture" as an inherited patrimony, a reified entity that can be maintained, lost, or recuperated.[10]

The third point to note is that Tukanoan ritual objects have a dual material-verbal character. Ritual objects are heirlooms both because they are inherited from previous generations and because they are intimately associated with an extensive verbal tradition. The objects are at once a material entity identified with an ancestor or ancestral body part and a set of ancestral words as verbal adjuncts. They have specific names, some of them the names of ancestors. They are associated with spells and ritual chants that refer to the human body as being made up of ritual objects and with narratives that rehearse in detail the objects' origins, various attributes, and ownership as property of a specific group. Ritual objects are therefore booklike in that they act as mnemonics that evoke the words with which they are imbued; in the case of flutes, they can also be made to speak, for flute music is the speech of the ancestors. As we have seen earlier, in addition to being read, Tibetan texts can also be spun around or "played" to make them speak.

The dual material-verbal character of Tukanoan ritual objects is also a feature of books themselves. Books are at once objects made of paper—Tukanoans called them *papera tuti*, "piles of paper"—and also "containers" of text as visible marks that evoke invisible, immaterial information: narratives, teachings, tradition, thought, knowledge, and so on. If anthropologists interested in literacy have tended to overlook the materiality of written documents, in Amerindian experience this materiality of documents is often as or more important than their textual content. This is not only because Amerindians were historically illiterate but also because, even today, the relatively few written documents that reach them are typically associated with powerful institutions and often have performative effects as instruments of power: Bibles and religious images, schoolbooks and calendars, letters patent and identity cards, money and traders' account books, land titles and legal documents. In Tukanoan speech, *papera* (paper) refers to both substance and concept in a way that is similar to our own notion of "papers" as official documents.

Given this experience, it comes as no surprise that, in some contexts, Tukanoans consider paper to be a potent, charged material and use it as a synecdoche for all of white people's power and technological know-how. They also draw analogies between literacy and shamanic knowledge and, more concretely, between white people's books and writing and their own feather ornaments and other sacred objects. These analogies emerge most clearly in versions of the near-universal Amerindian story of a fatal choice between two objects offered by a deity or culture hero that determines the respective powers and values of white people and Indians.

In addition to the usual choice between gun and bow (see Hugh-Jones 1988), some Tukanoan versions of this story bring paper and ritual objects together by presenting the fatal choice as one between feather ornaments (i.e., shamanic powers and knowledge) or guns and paper (i.e., different forms of foreign technical knowledge and power).[11] In another version of the story (Fulop 1954, 114), after giving sacred flutes and feathers ornaments to Indians, a cultural hero says: "You white people will have all the wealth in the world: arms, shotguns, knives, axes, notebooks, and papers. But you will never be able to remember things by memory. You will have to write them down on paper. Indians will not need to make notes on paper; they will remember all that happens by memory. And you white people won't be able to take away their memory" (translation mine).

Books and ritual objects thus emerge as two versions of the same thing: visible, emblematic manifestations of objectified "culture" that are also word objects, tangible things that condense intangible speech and knowledge. The story of a choice between books/paper and ceremonial goods can be read as a statement about cultural differences between white people and Indians as signaled by different kinds of goods. It can also be read as a statement that, in the new context in which the Tukanoans now find themselves, books are to be understood as *gaheuni*, ritual objects in a new, transformed form, objects that should now take their place alongside the more traditional feather ornaments, stools, gourds, gourd stands, and cigar holders.

This assimilation of books to ritual objects comes out most strikingly in a photograph of the ceremonial launch of one of the books in the NIRN series.[12] The origin story given in this book explains that the stacked objects illustrated on its cover, a cigar in a cigar holder supported in a gourd on a gourd stand itself standing on a stool, are the bodies of the deities who created the first human beings (see Hugh-Jones 2009). In the photograph, a Desana man dressed in feather ornaments holds a stool supporting a gourd stand that supports a copy of the book. The substitution of a book in place of the gourd, cigar, and cigar holder means that the book now instantiates in its physical form what it also tells as a story and shows as an illustration.

The context in which the book is being displayed, a ritual gathering to celebrate its publication, also suggests a further analogy, this time between the act of publishing and disseminating a book of sacred narratives and the ceremonial chanting of origin stories at *dabukuris* (ritual gatherings).

The books being published by the Tukanoans are thus hybrid forms, at once ancient heirlooms as words of their ancestors and also newly appropriated objects of modern, foreign power. Here they share some of the qualities of the axes, guns, and other potent objects that were appropriated in previous generations and assimilated to the indigenous category of *gaheuni* (sacred object, wealth). As compilations of sacred origin stories and other narratives, the message of the books lies in the text. But books as objects are also a potent attribute of outsiders that is now controlled and produced by Tukanoans themselves. Here the message lies as much in the material from which the books are made—in the paper itself.

**The Salesian Cultural Revolution and Its Aftermath**

For the peoples of the URN, initial contact with slavers and traders dates back to the early decades of the eighteenth century. Around the mid-nineteenth century, Catholic missionaries began operating in the area, and during most of the twentieth century, the Brazilian Uaupés was dominated by Salesian missionaries. The missionaries afforded a certain measure of protection from abuse by rubber gatherers and traders, but this protection came at the price of indigenous culture, in particular at the expense of the visible, tangible aspects of that culture. The Salesians tore down the Indians' elaborate communal houses, burned or stole their feather ornaments and ceremonial equipment, and moved the population into mission-dominated villages. This represented a conscious attack both on the foundation of the indigenous cosmos and on their collective rituals, feasting, and dancing, the *dabukuris* where different clans displayed their sacred objects and recited their origin narratives to make good their claims to prestige, status, and territory.

The missionaries also took the children away to boarding schools, where they were taught to read and write, forbidden to speak their own languages, and made to ridicule their traditional culture and way of life. Over the years, detailed knowledge of origin traditions, mythology, and shamanism became progressively confined to a dwindling generation of elders. As these elders died, their deaths were compared to the burning of a library.

Over the last decades of the past century, all of this began to change: with missionaries influenced by liberation theology, with constitutional reforms granting new political and cultural rights to indigenous peoples, and with NGOs linking ecological conservation with the conservation of indigenous cultures, the value accorded to indigenous culture underwent a rapid change. If, in the past, the missionaries had taught the Indians to ridicule their own culture as the antithesis of the civilization they should emulate, now culture was something to be cherished, reinforced, and recuperated. In this new global political context, "culture" is typically understood as the things emblematic of ethnic identity: language and mythology, traditional architecture, ritual, song, and dance. In lowland South America in particular, where groups such as the Kayapo use visual display and choreographed demonstrations to good effect in political meetings, recording them on video to be shown to a mass audience, feather ornaments and body paint have become the prime signals of indigenous identity (see Conklin 1997; Turner 1992).

In this new situation, the Tukanoans of the Brazilian Uaupés face a dilemma, for many of the visible signs of indigenous culture that would guarantee their authenticity as "real Indians" and have been used so effectively elsewhere in Brazil were long ago destroyed or taken away by the Salesians. In addition, the *dabukuris* that were the traditional arenas for displays of ritual objects and recitations of origin stories that underpinned claims to status and territory have declined and taken on new forms. It is with all of this in mind that a Tukanoan elder is reported as saying, "We have not lost all of our culture, only 50 percent" (Andrello 2006, 281). At the same time, with knowledge of origin stories, mythology, chanting, and shamanism largely confined to a few elder men, this other 50 percent, the verbal component of Tukanoan "culture," is also under threat.

The double character of books as at once both visible, material objects and containers of immaterial knowledge and verbal tradition makes them ideally suited as a vehicle through which to deal with this dilemma. On the one hand, they can preserve cultural traditions and transmit them to a literate younger generation who are well educated in the ideas and values of outsiders but relatively ignorant of their own traditions. On the other hand, they can render this verbal culture in the visible, tangible and potent medium of *papera*. The paper that Indians could once do without has now become both necessary and desirable.

To an external, nonindigenous audience, the Tukanoans' books direct a double message. They can be read as autoethnography or indigenous literature that affirms an identity that is at once generic as indigenous and specific as northwest Amazonian or Tukanoan—no other Amazonian peoples produce such books. As objects, they can also be "read" as signs of an indigenous modernity, of literacy, and of a system of education increasingly in the hands of indigenous communities. They thus form part of the politics of culture that is played out at national and international levels.

To an internal, indigenous audience, books also have a double quality. From them a younger, literate generation can learn the knowledge and oral traditions of their elders, thus ensuring that culture is not lost. At the same time, these traditions affirm the specific identity of the group in whose name to book is published and serve to validate the group's claims to status and territory. Cultural politics or the politics of publishing are here played out on a more local stage. A traditional mode of "publication," the display of sacred objects and the ritual chanting of origin narratives at *dabukuris*, is now complemented by another, the writing and publication of books as a new form of sacred object.

## A Choice of Technologies: Gun and Bow, Book or Video?

Thus far we have shown that for the Tukanoans, as for the Tibetans, books take their place alongside other sacred objects and ancestral relics, and like the digital archiving and cataloging going on in contemporary Tibet, the Tukanoans' enthusiasm for publishing is also related to their experience of religious persecution. But where does digital technology come in? Part of the answer is that it is precisely the digital technologies involved in sound recording, word processing, editing, and printing that have made the Tukanoans' publishing efforts possible.

Another answer might lie in the relations between digital technologies and the cultural features and current situations of the peoples who use them. For Tibet we have suggested that the way in which digital technologies are used in relation to books appears to be predicated on what books represent to Tibetans and to be associated with the effects of the Cultural Revolution. For the Tukanoans, we have suggested that a kind of elective affinity holds between books and traditional sacred objects and between their own forms of patrimony and wider notions of "culture" as patrimony, and books fit in with the narrative, verbal emphasis of Tukanoan culture that is under threat as a result of missionary activity.

Elsewhere Stephen Hugh-Jones (2010) has argued that there is a similar kind of affinity in the Kayapo's use of VCRs as an integral part of their choreographed politico-ritual demonstrations against hydroelectric dams and other threats to their survival. If the Tukanoans tend toward closure, ritual conservatism, and an emphasis on differences marked by the lineal inheritance of ancestral sacred objects and allied verbal traditions, the Kayapo are open, dynamic, innovative, and well practiced in acting in unison in newly created political rituals. Likewise if the Tukanoans assimilate books to *gaheuni*, a category of inherited verbal and material heirlooms, the Kayapo assimilate VCRs to *nekrêtch*, a category of ceremonial wealth that comes ideally as trophies captured from enemies (see Gordon 2006).

VCRs record primarily in a visual register with verbal content playing a secondary, supporting role, features that match the less verbal and more visual performative emphasis of Kayapo ritual displays. Books record primarily in a verbal register but are also visible, material entities as objects. These features fit in with the strongly commemorative verbal character of Tukanoan rituals and the assimilation of books to their sacred objects. In traditional Amerindian mythology, the choice was between a gun and a bow; today the choice is also between the book and the VCR. This is not a binary choice but more one of preference and of appropriate cultural fit.[13] Megaron Txucarramae, a Kayapo elder, is recorded as saying, "The video camera is like a gun, it is like a weapon"; Tolamān Kenhíri, a Tukanoan Desana elder, has written, "The book is the gun of the missionaries" (Tolamān Kenhíri 1993).

## Conclusion: The Many Lives of Books

What seemed originally to be a conventional operation of cataloging textual resources has turned out to be the starting point for a voyage of discovery exploring books as artifacts and technological innovation across very different cultures. Books and technological innovations proved to be so deeply embedded in their relevant contexts that they withstood any easy generalization based on the assumption that certain technologies are necessarily linked with the mind-set that generated them. The redefinition of communications technologies among Tibetans and Tukanoans shows that these have become means to appropriate and control the encounter with a modernity coming from elsewhere: digital technologies have come to Tibet via a secularist state, China, promoting a technological and scientific development expected to lead eventually to a disenchantment of the world, and yet they have become part of Buddhist rituality; the Tukanoans appropriate on their own terms the practice of publishing books, objects that were once used by missionaries to undermine Tukanoan culture; the Kayapo obtain VCRs from outsiders and use them as a propaganda weapon in their fight to preserve their lands and livelihoods from the depredations from those same outsiders. From this point of view, our cases all seem to be cases of the digital subversion addressed in general terms in the introduction. If similarities in the ritual significance of books suggest parallels between very different cultures such as the Tibetan and the Tukanoans, the difference between Tukanoans and Kayapo also shows that the relationship to technological innovation can vary considerably even among people living in the same general area. In this latter case, nuances of culture and different attitudes appear to favor the adoption of one form of communications media over another.

The cross-cultural look at books as objects with many experiential dimensions may also help in understanding their multilayeredness in the Western context. This multilayered quality, as well as the influence it has on the uptake of digital media in different contexts, suggests that nondigital and digital communications technologies more often than not coexist in different arrangements, thus questioning the common assumption that books will easily be substituted by digital surrogates. The stories of Tibetan and Tukanoan books percolating through international networks suggest that in a globalized world, books may well still have many different lives to live.

**Notes**

1. For an overview of the project, the content of the collection, and the events surrounding this operation of colonial collecting, see Diemberger and Hugh-Jones 2012.

2. This is a five-year collaborative project between the MIASU and the British Library involving also the Istituto Italiano per l'Africa e l'Oriente (IsiAO), the Nepal Research Centre, the Paltseg Research Institute, and the Tibetan Academy of Social Sciences in the Tibet Autonomous Region, as well as several other institutions across the world. It began in June 2010. The principal investigator is Uradyn Bulag, coinvestigators are Burkhard Quessel and Stephen Hugh-Jones, and the main research staff is Hildegard Diemberger.

3. Gell does not see persons as bounded biological organisms. Rather, "A person and a person's mind are not confined to particular spatio-temporal co-ordinates, but consist of a spread of biographical events, and memories of events, and a dispersed category of material objects, traces, and leavings, which can be attributed to a person and which ... may, indeed, prolong itself long after biological death" (Gell 1998, 122).

4. "The Tathāgata is your teacher, Ānanda. You have ministered to me, Ānanda, with friendly acts of body, acts of speech, acts of mind. Therefore then, Ānanda, just as you have given affection, faith and respect to me as I am at present in this incarnation, just so, Ānanda, should you act after my decease towards this perfection of wisdom. ... As long as this perfection of wisdom shall be observed in the world, one can be sure that 'for so long does the Tathāgata abide in it' that 'for so long does the

Tathāgata demonstrate dharma' and that 'the beings in it are not lacking in the vision of the Buddha, the hearing of the dharma, the attendance of the Samgha.' One should know that those beings are living in the presence of the Tathāgata who will hear this perfection of wisdom, take it up, study, spread, repeat and write it, and who will honour, revere, adore and worship it" (Conze [1973] 1994, 299–300).

5. One Tibetan commentary says: "Nagarjuna asked his benefactors, a Dharma king, to make a wheel containing the canon of Buddha's teachings to be turned." If this Tibetan text is correct, then the tradition of the revolving bookcase would have been initiated by Nagarjuna himself (Ladner 2000).

6. The underlying concept is that bones (Tib. *rus*) are transmitted from father to son. The spiritual abilities and the oral teachings of priests dealing with ancestral cults rooted in pre-Buddhist practices (Aya, Lhabon, etc.) are considered to be transmitted through the bones, and whenever these teachings were written down, they are passed on, as ritual objects, along the relevant bone line. The same idea has been applied to Buddhist tantric teachings whenever these are transmitted through lama lineages.

7. For more on this subject, see Hugh-Jones 2010 and Andrello 2010, two works that represent a collaborative endeavor.

8. *Gaheuni* is a Barasana word but has equivalent cognates in other Tukanoan languages.

9. See Hugh-Jones 2009.

10. "Culture" is placed in quotation marks to indicate a particular usage and understanding of culture as patrimony among lowland Amerindians and other indigenous peoples (see also Hugh-Jones 1997, 2010; Carneiro da Cunha 2009).

11. See Umúsin Panlõn Kumu and Tolamãn Kenhíri 1980, 73–74; Diakuru and Kisibi 1996, 175–176; Ñahuri and Kümarõ 2003, 205–207; Tõramü Bayar and Guahari Ye Ñi 2004, 230–232.

12. See Hugh-Jones 2010.

13. Tariana and Tukanoans from Yauarêtê (Uaupés, Brazil) have also been involved in a video project (see Carelli 2006), and others from the Pirá-Paraná area of the Colombian Vaupés are currently training in video techniques; the Kayapo and other Gê speakers have also published a book on the ecology and resources of their reservations (Programa de Formação de Professores Mêbêngôkre, Panará e Tapajúna 2007).

## References

Andrello, G. 2006. *Cidade do Índio: Transformações e cotidiano em Iauaretê*. São Paulo: Editora UNESP e ISA; Rio de Janeiro: NUTI.

Andrello, G. 2010. Falas, objetos e corpos: Autores indígenas no alto rio Negro. *Revista Brasileira de Ciencias Sociais* 25 (73): 5–26.

Barrett, T. 2008. *The Woman Who Discovered Printing*. New Haven, CT: Yale University Press.

Boutcher, W. 2012. L'object livre à laube de l'époque moderne. In *L'objet livre*, ed. S. Hugh-Jones and H. Diemberger. *Terrain* 59 (September): 90–103.

Carelli, V. 2006. *Iauaretê: Waterfall of the Jaguars*. Video. Olinda: Video in the Villages/INPHA.

Carneiro da Cunha, M. 2009. *"Culture" and Culture: Traditional Knowledge and Intellectual Rights*. Chicago: Prickly Paradigm Press.

Conklin, B. 1997. Body paint, feathers, and VCRs: Aesthetics and authenticity in Amazonian activism. *American Ethnologist* 24 (4): 711–737.

Conze, E. [1973] 1994. *The Perfection of Wisdom in Eight Thousand Lines and Its Verse Summary*. Delhi: Sri Satguru Publications.

Diakuru (Américo Castro Fernandes) and Kisibi (Dorvalino Moura Fernandes). 1996. *A mitologia sagrada dos Desana-Wari Dihputiro Põrã*. São Gabriel da Cachoeira: UNIRT/FOIRN.

Diemberger, H., and S. Hugh-Jones, eds. 2012. *The Younghusband "Mission" to Tibet*. Special issue, *Inner Asia* 14 (1).

Fulop, M. 1954. Aspectos de la cultura Tukana—mitología. *Revista Colombiana de Anthropología* 5:335–373.

Gell, A. 1998. *Art and Agency: An Anthropological Theory*. Oxford: Clarendon Press.

Goody, J. 1968. *Literacy in Traditional Societies*. Cambridge: Cambridge University Press.

Gordon, C. 2006. *Economia selvagem: Ritual e mercadoria entre os índios Xikrin-Mebêngôkre*. São Paulo: UNESP.

Hugh-Jones, S. 1988. The gun and the bow: Myths of white men and Indians. *L'Homme* 106–107:138–156.

Hugh-Jones, S. 1997. Éducation et culture: Réflexions sur certains développements dans la Région Colombienne du Pira-Parana. *Cahiers des Amériques Latines* 27:94–121.

Hugh-Jones, S. 2009. The fabricated body: Objects and ancestors in NW Amazonia. In *The Occult Life of Things*, ed. Fernando Santos-Granero. Tucson: University of Arizona Press.

Hugh-Jones, S. 2010. Entre l'image et l'ecrit, politique Tukano de patrimonialisation en Amazonie. *Cahiers des Amériques Latines* 63 (4): 195–227.

Ingold, T. 1968. *Lines*. London: Routledge.

Ladner, L. 2000. *Wheel of Great Compassion: The Practice of the Prayer Wheel in Tibetan Buddhism*. Boston: Wisdom Publications.

Maia, M., and Maia, T. 2004. *Üsâ Yēküsümia masîke: O conhecimento dos nossos antepassados; Uma narrativa oyé*. São Gabriel da Cachoeira: COIDI/FOIRN.

Āahuri (Miguel Azevedo) and Kümarō (Antenor Nascimento Azevedo). 2003. *Dahsea Hausirō Porā ukēsehe wiophesase merābueri tuti: Mitologia sagrada dos Tukano Hausirō Porā*. São Gabriel da Cachoeira: UNIRT/FOIRN.

Programa de Formação de Professores Mêbêngôkre, Panará e Tapajúna. 2007. *Atlas dos Territorios Mêbêngôkre, Panará e Tapajúna*.

Schaeffer, F. 2009. *The Culture of the Book in Tibet*. New York: Columbia University Press.

Strong, J. 2004. *Relics of the Buddha*. Princeton: Princeton University Press.

Tōramü Bayar (Wenceslau Sampaio Galvão), and Guahari Ye Ñi (Raimundo Castro Galvão). 2004. *Livro dos antigos Desana-Guahari Diputiro Porā*. São Gabriel da Cachoeira: ONIMRP/FOIRN.

Tolamān Kenhíri. 1993. La Bible et le fusil. In *Chroniques d'Une Conquête*. Special issue, *Ethnies: Droits de l'Homme et Peuples Autochtones* 14:19–21.

Turner, T. 1992. Defiant Images: The Kayapo appropriation of video. *Anthropology Today* 8 (6): 5–15.

Umúsin Panlōn Kumu (Firmiano Arantes Lana) and Tolamān Kenhíri (Luiz Gomes Lana). 1980. *Antes o mundo não existia: Mitologia dos antigos Desana-Këhíripōra*. São Gabriel da Cachoeira: UNIRT/FOIRN.

Warren, H. Clarke. 1984. *Buddhism in Translations*. New York: Atheneum.

# 6 Redeploying Technologies: ICT for Greater Agency and Capacity for Political Engagement in the Kelabit Highlands

Poline Bala

## Introduction

This chapter turns its gaze on the Kelabit in Central Borneo to highlight their ongoing engagement with information and communications technologies through the electronic Bario (eBario) initiative. eBario was a frontier ICT-based community development project implemented in 1999 to bring not only the Internet and computers but also the telephone to physically remote communities. Initiated as a partnership between researchers from Universiti Malaysia Sarawak (UNIMAS) and the Kelabit people of Bario, the project's main objective is to examine in what ways the new information and communications technologies (ICTs) such as the Internet, computers, printers, and VSATs can bring social and economic development to rural communities in Sarawak. It was initiated within the context of the ICT hype (Keniston 2002) of 1990s in which the new forms of ICT are seen as economic and social drivers that can further boost the economic and social development of societies. This has been substantiated by arguments that better telecommunications will induce rapid economic development (Barr 1998; Abdul-Rahman 1993).

Nevertheless, as the new technologies extend their reach, it raises questions of how the Kelabit in the highlands use or appropriate ICTs. Can engagement with ICTs change the ways in which communities like the Kelabit view themselves? Do ICTs give communities like the Kelabit hope and tangible evidence that external agencies are facilitating greater agency and self-advocacy? What happens to non-Western perspectives as the new technologies extend their reach? What is the impact of technologies on local notions of development and modernity? What do the Kelabit's experiences say about Malaysia's vision of a knowledge society, which reflects a messianic vision of societal transformation and economic progress through the embrace of technology? In what ways do Kelabit experiences disturb popular discourse in which technologies are perceived as "black boxes," or fixed entities that irresistibly change society and culture? To this end, is the assumption that technological developments are generally coercive structures more hindrance than help in analysis?

Based on data collected during twelve months of fieldwork (September 2005–September 2006), I aim to explore some of these questions. My argument is that it is the Kelabit's own desire for, and expectations of, "development" and "progress" that have conditioned their adopting and deployment of eBario. It is a quest that ties in closely with two fundamental Kelabit concepts: *doo*-ness (goodness) as social ideals and *iyuk*, or movement in social status among the Kelabit. I argue that the images and ideals of *doo*-ness, together with the interweaving processes between *doo*-ness and *iyuk*, have generated and sustained Kelabit modes of engagement and interaction with ideas, people, institutions, and objects from the "outside world." This includes their ongoing engagement with "development" and their recent responses to participative development, the Internet, telephone, and computers introduced through eBario. From this perspective, technologies do interact deeply with society and culture, but these interactions, whether in the form of resistance, accommodation, acceptance, and even enthusiasm, can be mediated and reconfigured by webs of social relations and the intricate interplay of social, political, and cultural conditions of specific society and culture. This is made clear in the way the Kelabit community in Bario engaged with the eBario Project.

The remainder of the chapter explores these various issues in more detail and is structured as follows. The next section highlights who the Kelabit people are, focusing especially on the social practices of traveling far and the bringing-in of novel objects, ideas, and technologies to the highlands—and how these relate to the fundamental concepts of *doo*-ness and *iyuk*. Both embody Kelabit images and ideals of *doo*-ness in terms of social honor and prestige in their society and goodness in conditions of life. Therefore, I suggest, an analysis from the perspective of practice theory (e.g., Ortner 1984) of how social honor and wealth are socially organized, and why they relate to images and ideals of goodness among the Kelabit, is necessary. Such an analysis can provide insight into, as Ortner has noted, "the context for understanding actors' motives, and the kinds of projects they construct for dealing with their situations" (152). Clammer (2001) also suggests that value constructing and generating activities can shape actors' responses to current changes in contemporary world.

This is followed by an exploration of what constitute "progress" and "development" for the Kelabit and what social and historical processes have given rise to their particular notion of development. Central to my argument is that it is through and within the traditional notions of *iyuk* and *doo*-ness that the Kelabit have internalized and engaged with new tropes of success and progress. Following on this argument, the third and final section will address the question of why the Kelabit wanted eBario to be brought into the Kelabit Highlands. I argue that the arrival of the Internet at Bario in the highlands represents a fulfillment of the Kelabit's own quest for development. As will be made clear by narrators in Bario, these technologies are both the means for, and signs of, what the Kelabit consider as *iyuk*, part of their continuous attempts and strategies to be associated with the larger world of progress (redeployment of technologies for greater agency).

**The Kelabit Situation: *Doo*-ness, *Iyuk*, and Objects of Desire**

The Kelabit population, numbering at 5,240 in 2000, traditionally lives in longhouses or villages dispersed across the remote, densely forested, hilly plateau of Central Borneo. Today these villages are Bario Lembaa, Pa' Umor, Pa' Ukat, Pa' Lungan, Pa' Mada, Pa' Dalih, Ramudu, Long Lellang, Long Seridan, Long Napir, and Long Peluan. Covering approximately 2,500 square kilometers with an average altitude of 1,000 meters above sea level, the plateau is located close to the international border between Kalimantan and Malaysia in Miri Division of the Malaysian state of Sarawak.

As usual in rural Borneo, the Kelabit historically were self-dependent or subsistence farmers: producing their own salt, planting their own rice, hunting, and gathering jungle produce for food and as a main source of protein and other nutrients. Although for years they have been far removed, in Tom Harrisson's words, "from what most people call 'the world'" (1959, 5), like many other Bornean societies, they have a long history of contact (or exchange, in the words Rousseau [1989]) and have shown a general inclination to appropriate the new and exogenous into their culture and ways of life. The assimilations of objects from outside are greatly facilitated by the existing cultural world of the Kelabit, which is pervaded by a concern for *iyuk* and *doo*-ness.

An ethnohistorical analysis reveals that it has been the images and ideals of *doo*-ness, together with the interweaving processes between *doo*-ness and *iyuk*, that have generated and sustained Kelabit modes of engagement with ideas, institutions, and objects from the outside world. While *iyuk* broadly refers to the notion of movement and specifically to status mobility, *doo*-ness embodies notions of goodness, success, and better well-being, or rather the qualities required to constitute a good person, such as knowledge, endurance and perseverance, self-discipline, hospitality, generosity, and strength. Attaining *doo*-ness and demonstrating these qualities associated with *doo*-ness underlie all social status among the Kelabit. In tandem with this, the Kelabit over the years have a created variety of strategies, activities, channels, and traditions through which to adopt and appropriate nonlocal or new ideas, people, and objects.

These are then "absorbed" into the Kelabit social systems and given new meanings to improve their well-being in the highlands. Examples are the adoption and appropriation of objects such as the *belanai ma'un* (old dragon jars), *ba'o* (beads), *tungul* (machetes), and *angai* (Chinese jugs).[1] Once these objects reach the longhouses, they are adapted into the Kelabit way of life—as visible signs of prestige in the community (Saging 1976/1977; Talla 1979; Janowski 1991).

One immanent and relevant activity through which objects of desire are absorbed into the Kelabit longhouses is the cultural practice of traveling far (*me ngerang mado*) to establish trade affiliations and build cultural contacts. It is arguably one of the most important strategies and traditions used by the Kelabit to forge connections with the rest of the world to attain *iyuk* and *doo*-ness. Through these trade affiliations, journeys, and cultural contacts, a diverse assortment of cultural borrowings and interactions have taken place between the Kelabit and the world outside, often obscuring the origin of these borrowings in the process (Bala 2002). This has included their ready embrace of Christianity, cash cropping, and also the kinds of material culture I discuss here.

**Shifting Contexts**

It is important to note that at the same time as the Kelabit absorb others, the Kelabit themselves, over the years, have been integrated to be part of a wider economic and political terrain. These simultaneous processes of assimilating and being assimilated have been long, dynamic, and complex and have great implications for the Kelabit social world, including what their notions of *iyuk* and *doo*-ness constitute in the contemporary world.

Turner's insight regarding performance and its transformative ability can perhaps contribute a further understanding: "To perform is to bring something about, to consummate something, or to carry out a play, order, or project. But in carrying it out ... something new may be generated. The performance transforms itself. The rules may frame the performance but the flow of action and interaction within that frame may conduce to hitherto unprecedented insights and even generate new symbols and meanings, which may be incorporated into subsequent performances. Traditional framings may have to be reframed—new bottles made for new wines" (Turner 1980, 160).

One area where Turner's analogy of new bottles for new wines is particularly revealing is the increasing importance of educational achievement, occupational prestige, wealth, and money among the Kelabit. The Kelabit today are eager to collaborate with the world of progress and development to attain better well-being (*doo ulun*) in the highlands. These new tropes of progress, success, and development are internalized as indicators of social status and prestige within the contemporary political and economic terrain. In a sense, they have also become the new contrastive measure by which the Kelabit define their local understanding of *doo*-ness and measure their own *iyuk*.

In this chapter, I highlight two situations that require the Kelabit to engage with new forms of defining *iyuk* and *doo*-ness, which emanate from beyond the local community. The first situation is the high level of rural–urban migration among the Kelabit, which has led to a geographically dispersed community. A survey conducted in 1998 revealed that 63.8 percent of the total Kelabit population has migrated out of the highlands (Murang 1998), and findings by Lee and Bahrin (1993) from ninety-three households in the highlands reveal that each household had at least one person who had migrated or moved away. They found that "about one-third had at least two members who had migrated; one-third had at least three–four members; and another one-third had between five and ten migrants from their household" (118). As argued by Amster (1998), large-scale population movement has colored the Kelabit social landscape today. In 2010, for instance, only 1,000 Kelabit remain in the highlands, while others have moved away for further education and better job opportunities in urban and town areas of Malaysia and beyond. These trends have not only transformed the Kelabit from a rice-farming-oriented community to one that produces professionals, religious leaders, and intellectuals who play important roles in the wider Malaysian society; they have also created a clear distinction between the Kelabit who remain in the highlands (rural Kelabit) and those who have left to live in urban areas (urban Kelabit).

The other situation is the incorporation of the Kelabit homeland into Malaysia in 1963.[2] The integration is a watershed in Kelabit social history, for it means engaging with Malaysia's comprehensive national planning for rapid economic progress and social change. As noted by Anderson ([1983] 1991, 12), "nation-ness is the most universal legitimate value in the political life of our time."

The interweaving of these two different yet interrelated occurrences had direct implications on the Kelabit's notions of *iyuk* and *doo*-ness. For instance, if in the past the Kelabit treasured and traveled in search of Chinese jars, beads, and other valuable items as part of their strategies for advancement, nowadays, of course, it is the pursuit of education, not barter trade and objects of desire, that is the main catalyst for the outward migration from the highlands (Murang 1998).

## The Kelabit and Malaysia's Planned Development

Scholars such as Robertson (1984), Hilley (2001), Larsen (1998), Scott (1985), King (1999), and Brosius (2003) share a common view that centrally planned development underpins Malaysia's policies, programs, and visions for modernization, providing a "vision discourse," or discourse of futurity, for the nation. Describing its dominance, Brosius (2003), after Bellah (1975, 3), suggests that the development paradigm in Malaysia today is a kind of "civil religion." An important dimension to Malaysia's national planning and policy is its ethnic framework of development. The policy signifies a development pattern, which is tilted toward allocation of scarce resources in the form of land, manpower, skills, capital, and development aid, albeit along racial lines. Conversely what this entails is identification of persons based on their ethnic and religious affiliation for the purpose of accessing political and economic resources in Malaysia (Jomo 1985). As noted by Chandra (1986, 33), ethnic and religious categories therefore "carry deep meanings for people" in Malaysia and concurrently force ethnic groups into a competitive relationships with each other, in which one group's advancement can mean another group's falling behind.

This competitiveness raises questions about how small communities like the Kelabit are integrated into the national quest for rapid economic growth. Although under article 153 (Kedit 1989) the Kelabit are "granted" Bumiputera (lit. the sons of the soil) status vis-à-vis non-Bumiputera (e.g., the Indians and Chinese), which guarantees the Kelabit certain privileges under the New Economic Policy (NEP), the eminent position of Malay *adat istiadat* and Islamic religion in the national agenda tends to draw the Malays close to the center of power, where they stand to get the most from the implementation of development under the NEP (Shamsul 1986). The sense of hegemony and superiority on the part of the Malays has often placed other ethnic and religious groups in a difficult position culturally.

For the Kelabit of Sarawak, their "peripheral situation" as "second-class Bumiputera" in relation to particular (Malay) political cultures is aggravated by the highlands' physical distance from centers of power. Without numbers, constituencies, pressure groups, or lobbies, and with their out-of-the-way location (Tsing 1993), there is a concern that the Kelabit are not heard in the context of a national integration discourse, which places the Malay-Muslim Bumiputera at the top of the hierarchy.

**The Kelabit Response**

The Kelabit have responded on two levels. The first level concerns a desire among the Kelabit as a minority group for collective political agency and social status within Malaysia's economic and political terrain. This is because a new form of *iyuk* competition has emerged, in which the Kelabit must engage competitively with other citizens who are not Kelabit for economic and political resources in the form of government financial support, government grants, development projects, and schemes (cf. Despres 1975, 2–3; Nagata 1979). Consequently a desire for what other ethnic groups have obtained and achieved because of their ethnic and religious backgrounds becomes a strong communal aspiration among the Kelabit in the highlands.

Conversely, the dynamics of the ethnic framework of development not only become a new contrastive measure by which the Kelabit define their local understanding of *doo*-ness and measure their own *iyuk*, but also provide a means to articulate and strengthen the Kelabit notions of ethnic identity and its boundaries (Barth 1969) and project their identity in relation to others in Malaysia and globally. The political features of Malaysia's ethnic framework of development has been, borrowing Goh's words, "reappropriated as idioms to derive power, class and cultural status from their positions within the state's modernizing discursive practices" (Goh 2002, 185).

The second level is the incorporation of the Malaysian government's development apparatus as a similar means for *iyuk* and another way of achieving *doo*-ness in the highlands. In this case, instead of voluntarily rejecting the injustice imposed by the hegemonic center, just as the Kelabit have incorporated jars, institutions, and other ideas as a means to enable *iyuk*, so the economic and social visions and strategies introduced by the government have been incorporated into the aspirations of those living in the highlands. The extent of this adoption can be gauged from the definition the Kelabit accord to the local notion of *iyuk* today, which is largely about development, in terms of getting or bringing progress to the villages through various development projects. In fact, to reject or even to question development is derided as not liking progress. A common phrase used to express this intermeshing of meaning is: "Without development we will not be able to *miyuk* [*Tulu naám pembangunan, naam teh tauh kereb miyuk*], and we will be left behind by the others [*ketedtan teh tauh let ngen lun beken*]."

*Iyuk* today has come to mean more: more security, economic growth, and employment; more pay, houses, and cars; and more and better education for their children. This is particularly prevalent among the younger generation, for whom being progressive and becoming modern, in their experience during the past thirty years, have been synonymous with economic boom. This is reflected in terms of lifestyle, educational achievement, and greater general prosperity among the Kelabit.

In other words, the Malaysian government's centrally planned development program has become a political and economic context within which concerns for status and prestige among the Kelabit are transformed to incorporate the desire for success and progress. The constitution of *doo*-ness has shifted: becoming modern has become the "appropriate" stage of being *doo*, associated with economic betterment, the acquisition of commodities and new technologies, and an affluent, successful, progressive and prestigious lifestyle.

The significance of all this is translated into the important roles of local power brokers, like the council of elders, the village headmen, the chiefs, the church council, the women's group, the educated elite, and the business class (e.g., shop and tourist lodge owners) in the highlands in the positioning and repositioning of Kelabit interests in relation to the state and the outside world. The aim is to ensure that development, in terms of funding and projects, do come to the Kelabit. Conversely, success in getting development, whether in the form of introducing new ideas or bringing in development projects and other resources to the highlands, has emerged as a political, economic, and social strategy to extend Kelabit collective agency within a wider political and economic context. These new things are all seen as means for enabling individuals to attain and enjoy an affluent, prestigious lifestyle, and at the same time for enabling the whole collective (the Kelabit society) to command high standards of living and respect from others within Malaysia's multiethnic setting. In fact, there is a sense of communal pride in being considered a progressive and successful community.

The interweaving of these aspirations and benefits has conditioned the Kelabit response to the idea of eBario in the Kelabit Highlands. As described earlier, the physical remoteness of the highland region means living with limited access to modern means of communication. As a result, the Kelabit desired a means to participate in an increasingly interconnected political and economic world. It was this desire, expressed as "wanting to be the same with the rest of the world and to be recognized as partners by others," that shaped their responses to the computers, telephone, and Internet provided by the eBario project. In line with their quest for progress, the Kelabit have adopted these new technologies to attain *doo*-ness and *iyuk* in the highlands, and likewise *doo*-ness and *iyuk* shape the way in which the Kelabit interpret or give meaning to these new technologies in their environment.

**What Is the eBario Project?**

Simply put, eBario is an offshoot of Malaysia's commitment to a centrally planned development for rapid economic development, which culminated in the institutionalization of Vision 2020, a road map for Malaysia to become a fully developed nation by the year 2020 (Mahathir M. 1991; Goh 2002, 190). The vision embodies the government's quest to be a part of the emerging knowledge-based global economy and society and ultimately the creation of a Malaysian knowledge-rich society.

In relation to this, the new information and communications technologies are seen as important keys to unlock opportunities and potentials to advance the process of development, and a platform needed to create a knowledge and information society (United Nations Economic and Social Council 2000, 9). Accordingly, ICT is aggressively being adopted for the creation of a Malaysian knowledge society.

As a result, one important aim is to expose every segment of Malaysian society to information and communications technologies (Larsen 1998; Goh 2002). This goal is being attempted through a collection of centralized, carefully planned and integrated initiatives. These include the Multimedia Super Corridor (MSC), which has seven flagship applications to demonstrate how ICT is expected to affect every Malaysian citizen's life and livelihood.

It was in response to this perceived need in national development planning that eBario was implemented as a pilot project to explore the economic, social, and cultural potentials of ICTs such as telephones, computers, very small aperture terminals (VSATs), and the Internet for rural development in Sarawak. Since access to ICTs is predicted to promote new social, economic, and cultural opportunities in rural areas (Ernberg 1998), eBario has grown to include the following physical and technological components. Computer laboratories were set up at the two schools in Bario, equipped with computers and printers. Telephonic equipment was installed within the existing communications network; for instance, telephones were placed at strategic locations or important meeting places in Bario, such as the airport, the shop area, the school, and the clinic. Very small aperture terminals (VSATs) and network configuration were installed by Telekom Malaysia Berhad, located at the shop area, the clinic, the school, and the airport. A telecenter known as Gatuman Bario (Bario Link) was set up in 2001 and equipped with computers, an inkjet printer, a laser printer, a laminating machine, a photocopier, and Internet access; since Bario is outside the national grid, the telecenter's alternative power supply was installed in the form of solar panels and diesel-run generators.

A IT literacy program was introduced in conjunction with COMServe, a local company based in Kuching. The training included word processing, keyboard usage, e-mailing, browsing the Web, and the management of technologies, including troubleshooting. Moreover, management and administration have been put in place to manage the project in Bario and also the community telecenter. To achieve this, a project coordinator-cum-manager has been appointed by the council of elders, JKKK, and UNIMAS to oversee the workings of the initiative in Bario. In addition to the project coordinator, a technical assistant was also trained and appointed to oversee the technical aspects of the project, such as troubleshooting and managing all the equipment and software.

### Redeployment of eBario in the Kelabit Highlands

At this point it is important to ask: in what ways have the Kelabit appropriated eBario as means to attain *doo*-ness and *iyuk* in the highlands and as a political, economic and social strategy to extend Kelabit collective agency within a wider political and economic context? To explore this, I turn to Miller and Slater's idea of the "dynamics of positioning" (2000, 18). It is a term used to describe people's ways of engaging with Internet media to position themselves within networks that transcend their immediate location, placing them within wider flows of cultural, political, and economic resources.

Among the Kelabit, these "dynamics of positioning" of eBario are obvious on two fronts. One is that it helps to ensure that the Kelabit in the highlands are not left behind but are on a par with the rest of the world (i.e., with other non-Kelabit group in Malaysia), by being connected to sophisticated modern means of communication and information exchange via the Internet and computer. The arrival of these objects and facilities in the Kelabit Highlands indicates that the villages have progressed: they are markers of collective *iyuk* and have placed them at the forefront of the more dynamic situations of competition for economic and politic resources in Malaysia.

Second is the way in which eBario helps and supports the community in relation to the far wider context of state and national development plans. In these situations, eBario has become both a forum and a stage, allowing the Kelabit to represent their aspirations for cultural and political recognition and for more development projects to be brought into the highlands.

### eBario as Marker of "Progress" and Success in the Kelabit Highlands

The first part of redeployment relates to the way that eBario is seen as *iyuk* and marker of progress and success among the Kelabit in Bario. All of this was made apparent by Tama Maren Ayu, the headman of Arur Dalan village and a prominent member of the council of elders. He said: "The arrival of these new technologies is an *iyuk* to us here in the highlands, similar to the arrival of chain saws many years ago. You see, before the era of chain saws, we were using ax (*kapek*), and before axes, it was the *uwai* (ax bow). *Uwai* was to us a prestige item; it came in through trading with people in the coast. Only those who were of high status (*doo*) had *uwai* as their possessions. These people were those who have endured the hardships of traveling far to the coast. Without *uwai* you couldn't make planks for your house, and as a result one could only build a small hut or even farmed on the hill slopes where the trees are smaller and easy to clear, but what you planted would not grow well. Without *uwai* your house would be smaller and thus couldn't house your relatives when they visited, your farm would be on the hill slope, and you would not have enough rice, which was important to feed others especially during feasts (*irau*). On the other hand, if a person had an *uwai*, he was able to *ngiyuk* (move) his social standings in the society."

Maren Ayu's remarks reflect that just as *uwai* was both a marker and an enabler of social status and mobility in the Kelabit village, so the new technologies are also seen to enable the Kelabit to move in their social standings in the highlands. It is a mark of their ability to attract progress to the Kelabit Highlands.

This perception was made more apparent through a dialogue session in Bario to discuss a site for the community telecenter. It was insisted that the new technologies be placed in a *new* building, not in the *old* airport terminal building, as suggested by the research team. This was because these machines were not merely sophisticated things for the Kelabit but also markers of their success. Balang Radu said: "We are the proud recipients of these new technologies in our area. It is similar to the arrival of the school in Bario in 1945. Tom Harrison came parachuting in. We didn't know who he was, but he brought us the school. We built new buildings for the school in those days. Just as now you have all arrived here with us, bringing in these new objects, we also need to build a new house for these new things that are arriving in our land." The insistence on a new building suggests a particular attitude toward modernity and development in the highlands: the Internet, computers, and printers were seen not merely as tools of development as proposed by the project team but also as signs and indicators of development arriving in the Kelabit homeland.

This view was shared by those who saw the adoption of the technologies in the Kelabit Highlands as empowering; it was considered a determined step toward charting a future for the highlands and Kelabit identity (Harris et al. 2001). This was based on the rationale that the project could improve the Kelabit economic and political position in Bario by acting creatively with development. In the words of Maren Tapan, "It is about moving in tandem with development, as opposed to being left behind by others [non-Kelabit]." Besides, according to Melayung Ulun of Bario Asal, "The Kelabit cannot afford not to adopt these new technologies. We need to be part of the change that is taking place; otherwise we will be lagging behind technologically and continue to find it difficult to communicate with the rest of the world." In the words of eighty-year-old Balang Radu, eBario has enabled further progress (*iyuk*) for those living in Bario by providing the means to forge connections with the rest of the world. He stated, "With these new means of communications, our lives are made much easier, although we live isolated in the headwaters of Baram. We can now liaise with the outside world from our villages, including talking to our children in KL, Kuching, and throughout the world. This is progress (*iyuk*) for us. It has made our life easier, and we are connected to the rest of the world in a new way. Therefore we are basically very, very pleased with its arrival. We are now on a par with the rest of the world."

Balang Radu's remarks demonstrate that the new technologies are being incorporated into the Kelabit's ongoing pursuit for *iyuk* in the highlands. This is attained by using eBario as a mechanism to position themselves within wider networks of interaction that transcend their isolated position in the highlands. In this way the Kelabit can *continue* to be integrated within (and be part of) the space of global flow of technologies, skills, communication, and information. As described earlier, Kelabit society has long been connected to the outside world through their geographic mobility and the dispersal of families. In tandem with their experiences, the Kelabit also see themselves as a part of the wider world of progress. Just as the cultural practices of traveling far and the adoption of objects and ideas have expanded their horizons, so too the contemporary acceptance of telephones, the Internet, VSATs, and computers in the highlands is seen as extending their existing connections to the rest of the world.

### eBario and the Maintenance of Kelabit *Doo*-ness (in the Form of Identity)

Being connected to the rest of the world through these new technologies is perceived not purely as a means of obtaining better-quality information, connectedness, and *iyuk* but also as a symbol that the Kelabit are not being left behind by others. In this context, eBario is taken as a signifier of recognition by higher and other agencies—researchers, government departments, and nongovernmental organizations at the national and international level. This is particularly significant within the Malaysian ethnic framework of development, in which ethnic groups have to compete for economic and political resources from the government, as discussed earlier. In other words, eBario represents a recognition of the Kelabit's distinct status as a people or ethnic group within Malaysia and within the international community. Seen in this light, the project is enabling, allowing the Kelabit to project their identity and status within Malaysia's multiethnic setting and within global flows of information and a larger world of progress.

This view was clearly expressed by the former paramount chief Pemanca Ngimet Ayu when addressing the council of elders, the team of researchers, and other Kelabit in June 1999. He said, "About thirty years ago Christianity, a new faith, came to us. We are now following the Lord. Our parents were very excited to embrace the new faith because it gave us better life. With the new faith, the school also came to us. We were also very excited to embrace the school because we foresaw that the school could help our children. Therefore we sacrificed to send our children to school. Today many of our children are doing well in school. In fact, many nowadays are in high positions. Again, today a new technology has arrived in our midst. Let's embrace it and see how we can use it in order not to be left behind by the rest of the world. We need to train our grandchildren to acquire these new skills and knowledge."

The paramount chief's statement reflects a general inclination among the Kelabit to explore computers and the Internet as a means of providing new skills and knowledge to the younger generation. These technologies are markers that the Kelabit are on a par with others in embracing worldwide shifts of perspective and influence. In other words, the arrival of these technologies in the highlands signifies not only that progress and development have arrived but, most importantly, that the Kelabit are active participants on a more global stage.

This sense of achievement is important for the communal status and collective interests in relation to others, especially within Malaysia's multiracial setting. As I have suggested, one of the ways in which the Kelabit engage with this is by positioning themselves on the same level as others in their pursuit of progress and success. With their capacity to attract new ideas and technologies into their environment, and their ability to adopt, incorporate, master, and re-create them, the Kelabit portray themselves as a successful and progressive people. eBario in this sense serves as a strategy for the image management of Kelabit *doo*-ness, in terms of their prestige and social status both locally and on the larger stage of Malaysia and the expanding world environment. It is a marker and signifier of Kelabit communal status, and a means of advancing Kelabit claims for access to more progress and development in the highlands. Although the Kelabit in Bario do not yet use the Internet to position their farming or hunting activities within wider farming networks, eBario as a developmental project has provided the means to advance their status and strengthen their identity in relation to a far wider context, and in a way that is much more dynamic than ever before.

The significance of this third aspect was made clear in a conversation with Gerawat Gala, the current president of Rurum Kelabit Sarawak. We were discussing how the media had misunderstood the meaning of the Top Seven Intelligent Communities Award given to the Kelabit in 2001. The media had reinterpreted the award as a tribute to the Kelabit for being a highly intelligent people based on their IQs. As the team leader, I was not particularly pleased with the media hype and its misrepresentations, but Mr. Gerawat's reply revealed a different perspective. "Let it be. It is a good thing for a small community like ours. We need the coverage of the project—both the image it created for us and the awards as a strategy for us—to encourage us to move on as a minority group, and also to position ourselves at par with the rest. The government can now witness that we are open to progress and are moving with the flow of development, and the others can see that we are not backward, but a progressive people."

Taking Gerawat's view a little further, good image management is of great importance to a community like the Kelabit, whose small numbers risk rendering them inconsequential in Malaysia's political processes and the global economy. I have described how NEP policies have shaped ethnic and religious categories as a means to access economic and political resources in Malaysia, and how this in turn is regulating interethnic relations between diverse ethnic groups, and between the society and the state in Malaysia. It is in relation to these competitive and ethnic conditions that eBario carries great significance for the Kelabit: the program represents international, national, and regional recognition of the Kelabit existence (peoplehood) and their active participation in the world of progress, rather than being seen as marginalized onlookers limited by their remote location and small numbers. It is a marker of the Kelabit's being at the forefront of competition for development in Malaysia, in which the presence of ICTs in the villages is associated with Kelabit openness and inclination toward progress and modernity. In other words, the Kelabit take pride in the project, for it promotes a notion that they are adventurous and willing to embrace opportunity and take risks, attributes that they wish others (non-Kelabit) would respect and emulate.

The last point is instantiated in the ways that the Kelabit are beginning to use ICTs to assert their cultural identity and their cultural rights. Like the Trinidadians observed by Miller and Slater (2000), the Kelabit have also become actors in global arenas through the Internet. The Kelabit are using the Internet to build their communal *doo*-ness and maintain communal networking and solidarity. One site that is used regularly by a number of the Kelabit community is the Online Kelabit Society (OKS). The significance of OKS in reproducing and maintaining solidarity among the Kelabit has been likened to the traditional roles of *ruma' kadang* (the longhouse) by one of its regular users. This is because it provides space for the traditional practice of social greeting or *peburi* and for the exchange of ideas and advice, which are important elements of communal living in a longhouse. As an online *ruma' kadang*, the site features discussions on various issues that face the Kelabit. One major topic or theme is Kelabit culture and identity: what does "being Kelabit" mean in a contemporary context, with high rates of rural–urban migration and high levels of intermarriage between Kelabit and non-Kelabit? This online forum and the discussions that take place within it allow for exchanges of ideas between members of the community both within Malaysia and beyond.

Other activities and pursuits include efforts to create family trees—not by one person but through group effort, as different members of the longhouses, now dispersed throughout Malaysia and also abroad, attempt to identify kinship relations between families. There are numerous online groups consisting of family members, close or distant cousins, and also special interest groups (e.g., golfers or those who are concerned about the encroachment of commercial logging and the impact of development in Bario). Other communal projects include the mapping of Native Customary Land and cultural sites in the Kelabit Highlands, and the documentation of the Kelabit language, tasks that are increasingly being managed via the Internet.[3]

The significance of the Internet is further illustrated by the ways in which eBario has been integrated within the local political apparatus to become a versatile platform for the Kelabit to position and reposition their interests within the contemporary political and economic terrain. The Internet, computers, and software are becoming useful tools and means to form networks, to acquire new skills, and to strategize the Kelabit's actions in their encounters particularly with new notions of development that include commercial logging and large-scale, futuristic development plans for the highlands. This differing concept of development has begun to shift attention away from socioeconomic development among the Kelabit to their legal rights and governance in relation to their land and cultural heritage in the highlands. This has stimulated individuals and groups to speak up after many years of moving in tandem with state-initiated plans for development.

The Kelabit have significant concerns about the potential impact of logging in the highlands. These include the effects on watersheds for wet rice cultivation, the Kelabit's dependence on the forest for jungle produce and wild game, and the rapid growth of ecotourism in the area. Many people provide guiding and lodging services for Malaysian and international tourists, many of whom are attracted by the opportunities for long-distance trekking. Seen in this light, demands for land for timber concessions are bound to come into conflict with highlanders who are competing for the same resources. This is because the very nature of logging is in complete contradiction to the new types of tourism that the Sarawak Tourism Board and the Kelabit themselves want to attract.

At the same time, there is a feeling that the Kelabit are dealing with the limitations of available local institutions and practices for confronting the many problems that commercial logging and road building are generating and will continue to generate. One critical issue is the shifting notion of land ownership, which on the ground is seen as a steady alienation of the Kelabit from their heritage land and, if left unaddressed, could become a growing arena for political conflict at the village level.

All these social and political processes are beginning to shape Kelabit modes of engagement with ICTs and their outcomes in the highlands. The use of the Internet, computers, and the telephone permits a form of political agency, especially as new forms of intervention threaten to change the physical and cultural landscape of the highlands. The new technologies inspire those in Bario to reach out to those who have left the highlands but still maintain a strong interest in the affairs of the village. These communications technologies have become a new means to maintain solidarity, within an increasingly stratified and occupationally mobile population, in the face of the new types of development intervention I have described. This suggests that ICT makes it possible for the Kelabit to form new networks and to reproduce effective organization and actions. At the same time, the presence of ICTs facilitates a greater agency and capacity for political engagement to question, assess, and debate these developments, and to form links with other agencies that might be useful.

Examples of this are the documentation of oral histories and the recording of images relating to all the cultural and historical sites found in the highlands, as well as the marking of their GPS points. These provide useful historical documents in negotiations with commercial logging as a new industry in the area. It is an example of local empowerment, whereby the use of ICTs has facilitated greater agency for political engagement in the face of this shifting notion of development. The new technologies are seen to increase the Kelabit's opportunities and abilities to make choices and to translate them into desired actions and outcomes.

Over the last ten years, the eBario project and in particular its telecenter has emerged to take on multiple meanings and functions in Bario. The project has become a symbolic compensation and a symbolic resource for the Kelabit's relative smallness in numbers, their political marginalization within Malaysia's ethnic framework of development, and the geographic isolation of the highlands from centers of power. At another level, it has become a meeting place, a community office, and a new platform to negotiate Kelabit interests and aspirations in the face of new notions of development being introduced in the highlands. Notably, the telecenter has been appropriated to facilitate greater agency and capacity for political engagement as the Kelabit engage with a growing tourism industry and the appearance of other large-scale commercial industries in the Kelabit Highlands.

## Conclusion

We can draw a number of significant conclusions from the Kelabit experiences with the eBario project. First, what the Kelabit say and think of development points to a particular shortcoming in the critiques of the development industry, which is the tendency to lose sight of the experiences and perceptions of the very people under study, particularly their definitions and understanding of development, and their reasons for desiring it. As has been argued, the Kelabit have incorporated the modern government apparatus as part of their strategies for *iyuk* (advancement) and as a means of connecting to, and collaborating with, the world of progress to attain better well-being (*doo ulun*) in the highlands.

In a sense, although they have been transformed in their meanings and forms, it is the continued significance of *iyuk* for status and *doo*-ness in the Kelabit social world that provides a rationale for the present-day attitude to change and the future, which in turn provides a basis for the adoption of ICTs in the social and economic environment in Bario. At the same time, the way in which the Kelabit provide meanings to the technologies deeply reflects how the Kelabit are eager to collaborate with the world of progress and development to attain better well-being (*doo ulun*) in the highlands and put themselves on a par with the rest of the world. This is particularly significant in the context of the Kelabit's marginal or even displaced position within the broader policies and discourse of the Malaysian state's ethnic framework of development. Although many do not use the Internet in relation to daily activities such as farming and so forth, the presence of the new technologies in Bario represents the Kelabit success in their pursuit of progress and development. It signifies that "we are at par with the rest of the world and people."

All of this resonates closely with Norman Long's transformative process of planned development, which he describes as "constantly reshaped by its own internal organization, cultural and political dynamics and by specific conditions it encounters and itself creates, including the responses and strategies of local groups who may struggle to define and defend their own social spaces, cultural boundaries and positions within the wider power field" (Long 2001, 72).

Moreover, ICTs as tools of development do not necessarily or simply open communities like the Kelabit to the outside world. Instead the Kelabit encounter with development as a hegemonic force closely resembles Michel de Certeau's understanding of agency "in exhibiting devious, dispersed, and subversive 'consumer practices,' which are not manifest through their own products, but rather their ways of using the products imposed by the dominant economic order" (1984, xiii). Placed within local social processes and circumstances, the new forms of technology as drivers of globalization have neither been used as a means of control (in the form of development as a tool of control by the state) nor heralded a new form of society (the creation of a knowledge-based society in the context of Malaysia). Rather, they have been partly integrated with, or subordinated to, existing practices, internal values, sociopolitical arrangements, and products in the community. Their continued use and adaptation have also provided for new forums of dialogue and communication, rekindling a sense of communal identity. They are therefore not separate entities but part of an ongoing and interactive process of change, which involves initiatives, responses, debate, occasional conflicts, and constructive resolutions. It is within these social processes that computers, the Internet, and the telephone have been given meaning, and their application modified and developed within the community's social context. In this sense, the visions of outside policy makers for introducing ICTs as tools for social and economic development may differ markedly from the actual realities of their use and effectiveness in different political and economic settings. ICTs are engaged with locally, interpreted, represented, and woven into the fabric of daily life of the Kelabit communities. As we can see, it is the Kelabit's own agency and skills that partly determine the impact and effects of the eBario project in the Kelabit Highlands.

## Notes

1. Many of these valuables were acquired from other traders like the Kayan, Kenyah, Berian, Potok, Kerayan, Murut, Malays, and the Chinese in return for tobacco, salt, rice, gutta-percha, and resin from the highlands. These dynamic relations of exchange crossed cultural and ethnic borders and were important not only for the Kelabit but also for the various groups in the interior, including the Malays and Dayaks (Rousseau 1989).

2. Elsewhere I have described how the process of this integration started with the expansion of the Sarawak territory beginning with James Brooke (1841–1868). However, it was only in 1920s and 1930s, through "headhunting raids," that the highlands were brought under Sarawak's rule. The Japanese occupation of 1943–1945 and the Indonesian-Malaysian confrontation in 1963–1966, in which the highlands became a center for military operations, have permanently rooted the Kelabit's position within Malaysia.

3. A recent example of this is an Internet forum to revive the use of the Kelabit language among the younger generation. The initiative was launched by a Kelabit woman living in Miri, who is concerned about the declining interest in, and use of, the language among migrant Kelabit. The Internet discussion list includes Kelabit who are living in Miri, Kuala Lumpur, Kuching, Bario, Bintulu, and Singapore. The main concern is to find ways of documenting "extinct" Kelabit words, terms, and phrases while at the same time promoting the use of the language.

## References

Abdul-Rahman, Omar. 1993. Industrial targets of Vision 2020: The science and technology perspective. In *Malaysia's Vision 2020: Understanding the Concept, Implications, and Challenges*, ed. Ahmad Sarji Abdul Hamid. Petaling Jaya: Pelanduk Publications.

Amster, Mathew H. 1998. Community, ethnicity, and modes of association among the Kelabit of Sarawak, East Malaysia. Ph.D. diss., Department of Anthropology, Brandeis University.

Anderson, Benedict. [1983] 1991. *Imagined Communities: Reflections on the Origins and Spread of Nationalism*. London: Verso.

Bala, P. 2002. *Changing Borders and Identities in the Kelabit Highlands: Anthropological Reflections on Growing Up in a Kelabit Village near the International Border*. Dayak Studies Contemporary Series no. 1. Institute of East Asian Studies, Universiti Malaysia Sarawak.

Barr, D. F. 1998. Integrated rural development through telecommunications. In *The First Mile of Connectivity: Advancing telecommunications for rural development through a participatory communication approach*, eds. D. Richardson and L. Paisley, 152-167. Rome, Italy: Food and Agricultural Organization of the United Nations. http://www.fao.org/docrep/x0295e/x0295e12.htm.

Barth, F. 1969. *Ethnic Groups and Boundaries: The Social Organization of Culture Difference*. Oslo: Universitetsforlaget.

Bellah, R. N. 1975. *The Broken Covenant: American Civil Religion in Time of Trial*. New York: Seabury Press.

Brosius, P. 2003. The forest and the nation: Negotiating citizenship in Sarawak, East Malaysia. In *Cultural Citizenship in Island Southeast Asia: Nation and Belonging in the Hinterland*, ed. R. Rosaldo, 76-133. Berkeley: University of California Press.

Chandra, M. 1986. Territorial integration: A personal view. In *The Bonding of a Nation: Federalism and Territorial Integration in Malaysia*. Proceedings of the First ISIS Conference on National Integration, Kuala Lumpur, October 31–November 3, 1985. Malaysia: ISIS.

Clammer, J. 2001. *Values and Development in Southeast Asia*. Subang Jaya, Malaysia: Pelanduk Publications.

de Certeau, M. 1984. *The Practice of Everyday Life*. Berkeley: University of California Press.

Despres, L., ed. 1975. *Ethnicity and Resource Competition in Plural Society*. The Hague: Mouton.

Ernberg, J. 1998. Empowering communities in the information society: An international perspective. In *The First Mile of Connectivity: Advancing Telecommunications for Rural Development through a Participatory Communication Approach*, ed. D. Richardson and L. Paisley, 191–211. Rome: Food and Agricultural Organization of the United Nations. http://www.fao.org/docrep/x0295e/x0295e15.htm.

Goh, B. L. 2002. Rethinking modernity: State, ethnicity, and class in the forging of a modern urban Malaysia. In *Local Cultures and the New Asia: The Society, Culture, and Capitalism in Southeast Asia*, ed. C. J. W.-L. Wee, 184-216. Singapore: Institute of Southeast Asian Studies.

Harris, R. W., P. Bala, P. Songan, and G. L. Khoo. 2001. Challenges and opportunities in introducing information and communication technologies to the Kelabit community of North Central Borneo. *New Media and Society* 3 (3):271–296.

Harrisson, T. 1959. *World Within: A Borneo Story*. Singapore: Oxford University Press.

Hilley, J. 2001. *Malaysia: Mahathirism, Hegemony, and the New Opposition*. London: Zed Books.

Janowski, M. 1991. Rice, work, and community among the Kelabits in Sarawak, East Malaysia. Ph.D. thesis, London School of Economics, University of London.

Janowski, M. 2003. The forest, source of life: The Kelabit of Sarawak. British Museum Occasional Paper 143. The Sarawak Museum.

Jomo, K. S. 1985. *Malaysia's New Economic Policies: Evaluations of the Mid-term Review of the 4th Mp*. Kuala Lumpur: Malaysian Economic Association.

Kato, Tsuyoshi. 1982. *Matriliny and Migration: Evolving Minangkabau Traditions in Indonesia*. Ithaca, NY: Cornell University Press.

Kedit, P. M. 1989. Ethnicity in multi-cultural society: Dayak ethnicity in the context of Malaysian multi-cultural society. *Sarawak Museum Journal* 40 (61):1–7.

Keniston, K. 2002. IT for the common man: Lessons from India. M. N. Srinivas Memorial Lecture, Indian Institute of Science. Bangalore: NIAS Special Publication SP7-02.

King, V. T. 1999. *Anthropology and Development in South-East Asia: Theory and Practice*. Kuala Lumpur: Oxford University Press.

Larsen, A. K. 1998. Discourses on development in Malaysia. In *Anthropological Perspectives on Local Development: Knowledge and Sentiments in Conflict*, eds. S. Abram and J. Waldren, 18-35. London: Routledge.

Lee, B. T., and Tengku Shamsul Bahrin. 1993. The Bario exodus: A conception of Sarawak urbanization. *Borneo Review* 4 (2): 112–127.

Long, N. 2001. *Development Sociology: Actor Perspectives*. London: Routledge.

Mahathir, M. 1991. The way forward. Working paper presented at the Malaysian Business Council, February. http://www.pmo.gov.my/?menu=page&page=1904.

Miller, D., and D. Slater. 2000. *The Internet: An Ethnographic Approach*. Oxford: Berg.

Murang, O. 1998. Migration in Sarawak: The Kelabit experience. Presented at Workshop on Migration in Sarawak, organized by the Sarawak Development Institute, June 25–26, 1998, Parkcity Beverly Hotel, Bintulu.

Nagata, J. 1979. *Malaysian Mosaic Perspective from a Poly-ethnic Society*. Vancouver: University of British Columbia Press.

Ortner, S. B. 1984. Theory in anthropology since the sixties. *Comparative Studies in Society and History* 26:126–166.

Robertson, A. F. 1984. *People and the State: An Anthropology of Planned Development*. Cambridge: Cambridge University Press.

Rousseau, J. 1989. Central Borneo and its relations with coastal Malay sultanates. In *Outwitting the State*, ed. P. Skalnik, 41-50. New Brunswick, NJ: Transaction Publishers.

Rousseau, J. 1990. *Central Borneo: Ethnic Identity and Social Life in a Stratified Society*. Oxford: Clarendon Press.

Saging, R. 1976/1977. An ethno-history of the Kelabit tribe of Sarawak: A brief look at the Kelabit tribe before World War II and after. Graduation exercise submitted to the Jabatan Sejarah, University of Malaya, in partial fulfillment of the requirements for the degree of Bachelor of Arts.

Saging, R., and L. Bulan. 1989. Kelabit ethnography: A brief report. *Sarawak Museum Journal* 40 (61):89–118.

Scott, J. C. 1985. *Weapons of the Weak: Everyday Forms of Peasant Resistance*. New Haven, CT: Yale University Press.

Shamsul, A. B. 1986. *From British to Bumiputera Rule: Local Politics and Rural Development in Malaysia*. Singapore: Institute of Southeast Asian Studies.

Talla, Y. 1979. *The Kelabit of the Kelabit Highlands, Sarawak*. Ed. Clifford Sather. Report no. 9, Social Anthropology Section, School of Comparative Social Sciences, USM, Pulau Pinang.

Tsing, A. L. 1993. *In the Realm of the Diamond Queen: Marginality in an Out-of-the-Way Place*. Princeton, NJ: Princeton University Press.

Turner, V. 1980. Social dramas and stories about them. *Critical Inquiry* 7 (1): 141–168.

United Nations Economic and Social Council. 2000. Draft Ministerial Declaration of the High-Level Segment Submitted by the President of the Economic and Social Council on the Basis of Informal Consultations, New York.

# 7 Making the Invisible Visible: Designing Technology for Nonliterate Hunter-Gatherers

Jerome Lewis

## The Invisible People

Pygmy hunter-gatherers in the Congo Basin are reputed simply to vanish when danger is imminent. They are famous for supposedly using such skills when accomplishing seemingly impossible tasks such as single-handedly spearing a fierce forest elephant or vanishing suddenly into the forest and then reappearing just as suddenly when it suits them. Their fearfulness of incoming agriculturalists and fisher groups led them to avoid such newcomers, sometimes for many years, before tentatively making contact through practices such as silent trade.[1] One consequence of this is that non-Pygmy groups in the Congo Basin frequently call Pygmy hunter-gatherers "Twa," "Tua," or "Cwa" and other similar ethnonyms that translate as "the invisible people" (Schadeberg 1999).

While this reputation for invisibility deservedly lives on, in this chapter I focus on how it has contributed to their being ignored in land management decisions that concern forest they depend on. Today governments and outsiders such as loggers or conservationists can exploit Global Positioning System (GPS) technology to find their way in and out of forest areas to which access was once controlled by Pygmies, and geographic information systems (GIS) facilitate urban, office-based land management decision making. Supported by these technologies, nonforest people are able to present the forest in ways that privilege their interests: governments as providing development and tax-generating opportunities, loggers as containing rich sources of high-value timber, and conservationists as biodiverse environments deserving protection from people. These presentations of value, often put forward using maps that do not acknowledge the presence of Pygmies, lead to decisions that can have profoundly negative consequences for forest people as they are steadily disenfranchised of land and resources. Their forest occupation, management practices, key resources, and livelihood needs have been ignored, at least until very recently.

This chapter explains how they are now making their own maps using modified hardware and special software. These maps become their emissaries, able to communicate their most pressing concerns to powerful outsiders in office-based meetings to which Pygmies would never be invited. Indeed, maps have proved far more effective than more traditional advocacy methods based on meetings, workshops, and research papers. To make these maps, hunter-gatherers have learned new techniques to represent their forest use and spiritual values and to monitor other people's use of the forest. Although the map-making project is still in its infancy, it is hoped that these solutions will eventually become available to increasing numbers of forest people and encompass ever greater areas of forest.

Developing these solutions depended on questioning certain presuppositions of software developers while at the same time collaborating with forest people who are highly skilled and knowledgeable but mostly nonliterate, seminomadic, and technologically minimalist. This chapter describes the participatory design process that is spawning local modifications to the software so that it can be redeployed to address new problems. This trajectory of development is promoting rapid technological leapfrogging among participating forest communities. Now that the forest people are able to make their presence visible and to communicate their perceptions of value to outsiders, for the first time their concerns and needs are being taken into account in forest management decisions.

While the outcome of these technologies is a greater voice for forest people in forest management, it avoids risks that they associate with overt political engagement or representation. As I shall explain, these risks are at once to their egalitarian political organization, which structurally resists the emergence of leaders or representatives, and from dangers they connect with appearing to resist powerful outsiders' desire to obtain wealth. In the context of regional conflicts between bigmen for control of resources, anonymity and political neutrality are greatly valued by participating hunter-gatherers.

This chapter focuses on work with Yaka and Baka hunter-gatherers in Congo-Brazzaville and Cameroon. They have been the main partners in developing and using these new mapping solutions. Later developments also include neighboring farming groups who have joined in redeploying the technology to monitor resource extraction in Cameroonian forest.

### Some Technological Assumptions That Matter

As a casual glance at most software interfaces suggests, there is a basic assumption that users are competently literate (often in English):[2] it underpins how developers expect users to interact with the devices on which their software will run. The assumption is largely justified, since school education has become a rite of passage for most of the world's children. Nonliteracy is not normally a consideration for designers.

Though some nonliterate people exist everywhere, they constitute the overwhelming majority in places like the forests of the Congo Basin. Among the total population of approximately fifteen thousand Yaka hunter-gatherers living in the area in which I have worked since 1994, I have met only a handful of men with basic literacy skills. The problems of assuming general literacy in software design are thrown into relief by the vital need to provide accessible technology so that such marginalized people can really make their voices heard by powerful others. Encouraging literacy skills is the obvious long-term solution, but obtaining literacy is not quick, nor is it a value-free process.

Since the late colonial period, schools have been a key site for indoctrinating children into the values of culturally dominant elites. Despite independence in Central Africa, schools continue to provide this function. In the case of Pygmies, until now schooling and obligatory sedentarization continue to be the most commonly implemented government strategy to enforce assimilation. In this context, as in Europe and elsewhere in the world, primary and secondary schooling serves to mold future citizens into conformity by training children in disciplines suited to complex state management. These include imposing punctuality, particular bodily comportments, and styles of self-presentation; defining success in terms of conformity to accepted bodies of knowledge and unquestioned acceptance of authority; and instilling the values of competitive individualism through grading and examinations. These practices and values frequently contrast with, and even contradict, more local learning styles and values.

The key learning contexts for egalitarian Yaka hunter-gatherers are not structured around the hierarchical teacher-pupil relationship. Rather, they promote and emphasize lifelong, self-motivated, cooperative group learning through mimicry of those slightly more competent than oneself. In this peer-to-peer learning, participation is stimulated by enjoyment, curiosity, or excitement, and success is achieved by cooperating with peers, not by competing with them. These nonhierarchical learning methods support the Yaka's egalitarian political order while ensuring that they cultivate the refined cooperative and mimicry skills they need to become proficient forest hunter-gatherers (Lewis 2002, 124–197). Sensitive group coordination in conjunction with individual expertise is the key to successful forest living. They learn all the key skills they need without dependency on specific others and so maintain their egalitarian social and political order (Lewis 2008a). Although extremely capable at reading each other, tracks, and other forest signs, they do not learn to read print.

Now, widespread nonliteracy among Congo Basin hunter-gatherers puts them at a disadvantage when trying to get their rights respected. State governments ignore their land rights, auctioning off their resources to multinational industrial corporations and international conservation organizations. Having obtained state authorization, these powerful outsider organizations can then exploit national forces of law and order to impose their agenda on forest areas that the hunter-gatherers had always considered their own and on which they depend for their livelihood.

**A Disguised Colonialism: Roads**

In the two-million-hectare area that I have been working in since 1994,[3] about 15,000 Yaka associate themselves with specific forest areas varying in size from 150,000 hectares to 500,000 hectares. They require these relatively large areas to gain a livelihood sustainably and reliably. Within these areas, people know the resources intimately and walk between them according to season and desire. During three years of doctoral fieldwork, I walked around 6,000 kilometers. I remember the first time the people I lived with saw a newly completed road. How they complained about the hard laterite surface hurting their feet! When the first logging truck rumbled toward us, all the women and children fled, as if from a charging buffalo, twenty meters into the undergrowth on either side of the road.

Since that time in 1995, high prices for tropical timber have fueled a massive expansion of roads into remote forest areas in search of valuable trees. Urban populations come together rapidly in remote forest locations, attracted by the job opportunities offered as industry expands. Small towns grow around the activities of the logging company. Traders and a range of other service providers come to support the workers. The local logging town in my fieldwork area of 6,000 people in the late 1990s had become a town of around 15,000 people by 2007.

These developments have wrought great changes on the Yaka's forest. Roads have multiplied and spread out to open up previously inaccessible areas to outsiders who can now extract valuable resources in huge amounts. Suddenly resources the Yaka thought they managed—such as the wild animals—were up for grabs. After the mid-1990s, commercial bush meat hunting became a serious problem in certain areas, notably in forest where new roads were being cut. In response, Euro-American conservationists set up paramilitary patrols with the intention of catching poachers.

While this has resulted in limited successes in certain cases, it has also caused serious problems for many Yaka. The distinction between a Yaka hunter and a poacher is often ignored. Eco-guards frequently make them scapegoats in their antipoaching controls. Since the most serious environmental crimes are mostly organized by local political and military elites, eco-guards are often unable to arrest perpetrators. So Yaka become soft targets in their forest camps for violent visitations by these paramilitary groups. The beatings and other abuses experienced during these visits are a source of great anger among the Yaka (see Lewis 2008b; N'zobo et al. 2004).

Since time immemorial, avoidance has been an effective strategy used by hunter-gatherers to distance themselves from others in their forest. Often newcomers do not even realize that the hunter-gatherers are there. With such an insignificant impact on the environment, it is easy for them to remain invisible for extended periods. Pygmies' oral history attests to their shock at witnessing brutal intervillage raiding and warfare in the precolonial period, or colonial violence against village populations during forced-labor regimes, and more recently the actions of militias targeting civilians during intermittent civil war in the 1990s. These traumatic events have resulted in them avoiding or being extremely wary of landscapes outside the forest, whether riverbanks, villages, or towns.

Previously, by moving back into deeper forest, they could avoid danger and continue to live autonomously. Their very low population density—averaging about 0.2 persons per square kilometer over the whole Yaka area—enabled avoidance to be effective. Now, with the widespread penetration of logging roads and the patrols of eco-guards, Yaka tell me that they feel less safe in small forest camps. Women especially fear raids by the eco-guards. One response has been to live in much larger forest camps, sometimes reaching over one hundred members. These super-camps are characterized by fluid composition and great instability caused by social tensions, contagious illness, and hunger. They regularly break up due to conflict, then disperse and reform elsewhere. Hunting and gathering works best in much smaller groups.

As more and more parts of the forest become dangerous for Yaka, avoidance has become less effective. As new roads open up new parts of the forest to outsiders, so too they draw new groups of Yaka out to the logging town to "open their eyes" (*dibua miso*). To open one's eyes means to become aware of where manufactured goods come from (the market stalls), how much they cost, to see the white man's magic (machines, especially airplanes), and how many Yaka and other sorts of people there are in the world.

Their behavior in this new environment reminded me of that of tourists. They would spend a great deal of time hanging around the big machinery or the market, carefully observing comings and goings. Some would take jobs only to leave before they had received their first pay packet. To my initial surprise, rather than cash or shotguns, they would take new rituals and songs as souvenirs to share back home. Without numeracy or an understanding of money, they invariably suffered, and so the logging towns earned the name *ebende ya njala* (place of hunger), and most people would not stay for more than a few months.

**The Role of Maps**

As I watched these processes unfolding, I thought there was nothing I could do apart from documenting and reflecting on them. In particular, I noticed the continuity between the role that maps played in enabling outsiders to divide up the forest and to organize their activities regardless of forest people, both during the colonial period and since independence. In the last ten years, improvements in mapping technology have facilitated an unprecedented microcontrol of forest resources that was never possible in the past.

Although earlier maps existed, the most accurate and useful were a set of 1:200,000 scale maps covering French Equatorial Africa produced in the 1950s. These maps are visually dominated by large swathes of green "uninhabited" primary forest on firm land surrounded by a patchwork of marshy and semiflooded forest, occasionally broken by clearings or veinlike streams and rivers. Since the land is relatively flat, there are hardly any contours to use in orientation. While certain international boundaries were decided in Europe using a ruler placed over a map, the boundaries of all modern forestry concessions carefully follow the limits of the forest areas on firm land. Large marshes or areas of flooded forest do not produce commercially exploitable trees and are excluded.

Modeled on the colonial system, the Congolese government's delimitation of the forest into large concessions called UFAs (Unitie Forestieres d'Amenagement, or Forest Management Units) was intended to attract investment from foreign companies wishing to exploit forest resources. Two major bursts of activity seeking companies to exploit these concessions occurred first in the late 1970s and early 1980s, and then in the late 1990s and early 2000s. In the first period, the focus was on a few concessions with river access. Owing to the lack of roads or efficient road-making

# Making the Invisible Visible

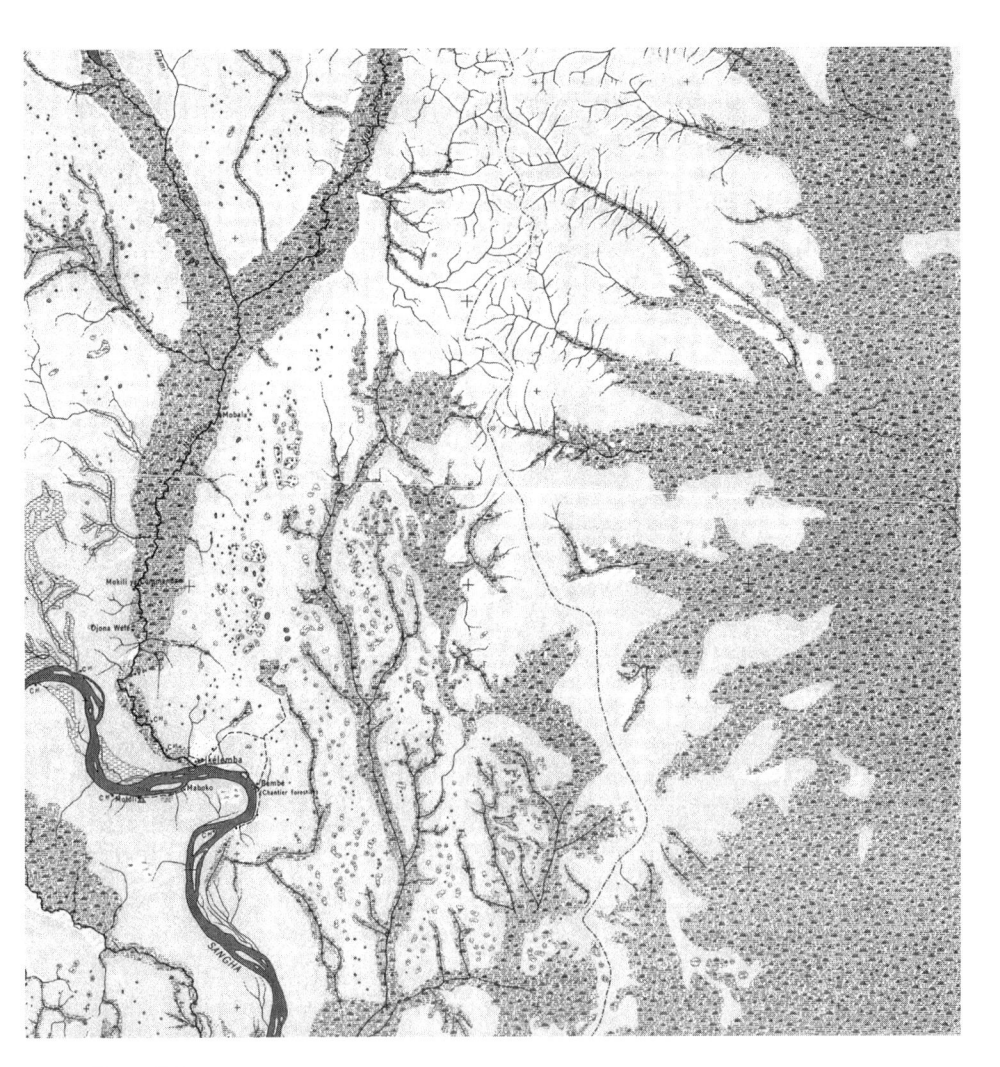

**Figure 7.1**
Map of Yaka forest, northern Congo-Brazzaville. Scale is 1:200,000.

machinery, logs had to be dragged by tractors to the riverbank and floated down toward the sea.

In the second burst of occupation, the entire forest was included as new advances in both hardware (GPS machines, powerful earthmoving and road-making gear) and software (GIS and a variety of forestry-specific programs relating to road planning, felling, and transformation procedures) made remote forest commercially viable for logging. Today's loggers can

134　　　　　　　　　　　　　　　　　　　　　　　　　　　Jerome Lewis

## Atlas Forestier de la République du Congo

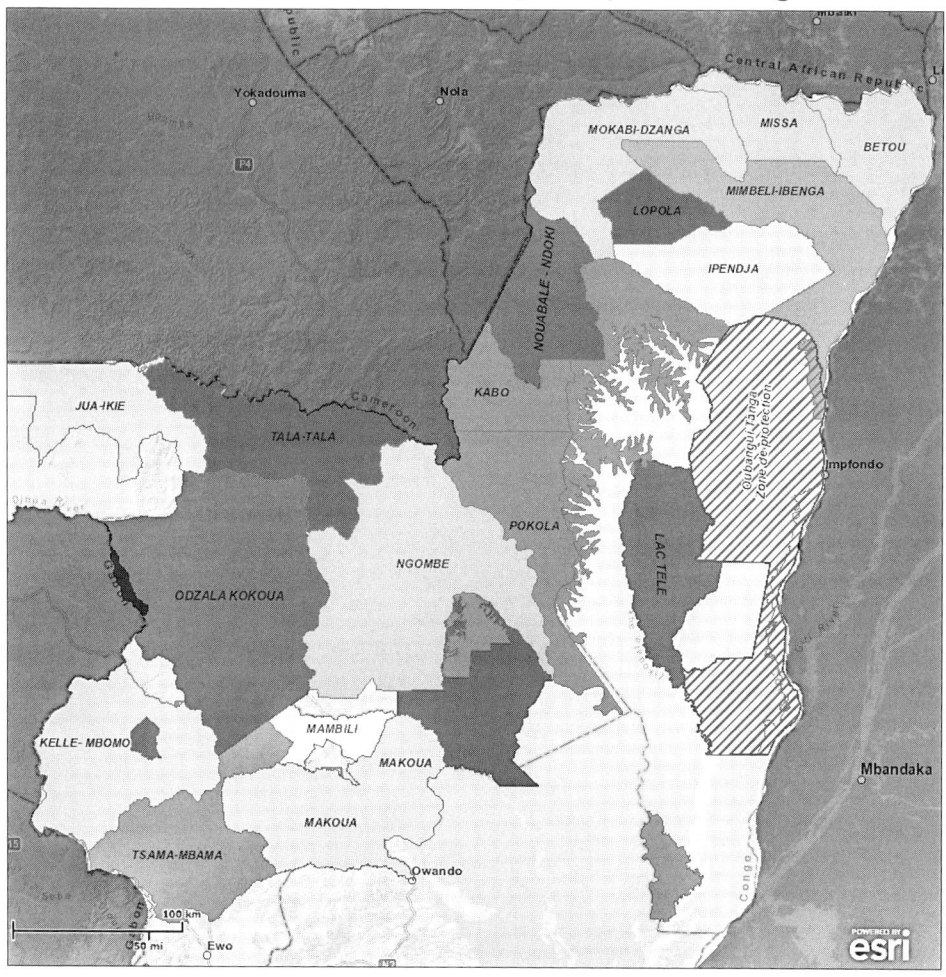

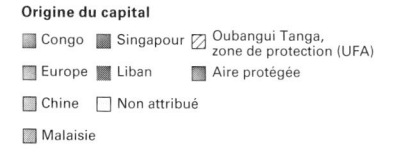

**Origine du capital**

- Congo
- Singapour
- Oubangui Tanga, zone de protection (UFA)
- Europe
- Liban
- Aire protégée
- Chine
- Non attribué
- Malaisie

La carte de l'origine du capital montre le pays d'origine des investisseurs primaires des concessions forestières en République du Congo.

Fri Sep 6 2013 12:03:02 PM

WORLD RESOURCES INSTITUTE

**Figure 7.2**
Concessions in northern Congo-Brazzaville.

quickly overcome major obstacles to build high-quality roads into any part of the forest. These roads make possible the independent exploration of remote forest areas using GPS technology, which prevents workers from getting lost. Prospectors can then log their findings and organize exploitation independent of local people using GIS platforms.

The green swathes encompassed by concessions on the maps represent forest inhabited by Yaka. Yaka divide the green into individual areas of forest bounded by streams or marshes. Each one is the responsibility of a named clan that attends to its long-term well-being by visiting it and sharing in its resources while also performing rituals that ensure its continued abundance. While this intimate and detailed knowledge of the landscape attests to a long and enduring relationship with the forest, it lacks representation. In contrast, forest temporarily cleared for farming or villages is marked and often named. This reflects national laws that define landownership in terms of the land's visible transformation. This discriminates against Yaka land use and claims over land, since they do not transform their environment to use it, and the majority of their lands will appear unoccupied at any given time.

The comprehensiveness of forest covered in the latest round of delimiting and renting out concessions was officially justified as part of postwar reconstruction to promote local development. All the concessions were rented out to logging companies or attributed to conservation organizations. Other outsiders, mostly commercial hunters supplying urban centers with bush meat, use the loggers' infrastructure to exploit previously inaccessible areas. The impact of these various uses is that local people—both Yaka and villagers—see their resource base diminishing and increasing numbers of strangers coming into their forest. In practical terms for local people, their forests have been converted into faunal and floral assets and rented out by the central government.

Since the central government ignored the hunter-gatherers living in the forest, so too did the logging companies and wildlife conservationists. Which trees would be cut down were determined by the loggers and approved in the capital, as were park boundaries drawn by the conservationists. Indeed, until very recently, unless local people caused problems to these activities, they were simply ignored.

**Forest Stewardship Council Certification and Responsible Forestry**

However, mounting criticism of the tropical logging industry in the context of deforestation, climate change, and the huge profits being made despite the increasingly impoverished status of forest people led to growing calls for responsible and more equitable forestry practices. The most widely acceptable standard with which to judge the achievement of responsible environmental and social forestry is the Forest Stewardship Council (FSC) certificate. In the early 2000s, certain leading forestry companies in the Congo Basin began to work seriously at addressing the principles and criteria of the FSC in their operations.

To address the environmental side, they began studying the natural regeneration rates of the key commercial species, mapping their distribution in age groups across their concessions and calculating sustainable harvests and rotations. They began experimenting with replanting strategies and developed partnerships with conservationists to manage wildlife in their concessions. They sought to reduce the collateral damage to the forest by improving felling techniques, reducing road widths, and attempting to control outsiders' access to the forest. This has required huge investments by the logging companies over many years, but they are slowly beginning to define what responsible forestry in the forests of the Congo Basin should entail.

While these advances in environmental sustainability are significant, much less progress has been made to address the social aspects of the certificate's principles. Although improving safety and working conditions for staff has been achieved, the requirements relating to forest inhabitants have been the subject of much controversy and highly variable application. Principles 2 and 3 of FSC oblige companies to respect the rights and resources of indigenous and other forest people on whose land forestry activities occur; this may range from obtaining their free, prior, and informed consent to involving them in forest management decision making when it concerns forest they depend on.

Principle 3, for instance, "Indigenous People's Rights," is meant to ensure that "the legal and customary rights of indigenous peoples to own, use and manage their lands, territories, and resources shall be recognised and respected." More controversy surrounds the application of this FSC principle than any other. To avoid this, one large FSC-certified company categorically refuses to work in forest inhabited by indigenous people. Auditors have been unable to apply the same standards in assessing compliance to Principle 3, and so a number of certificates awarded in the Congo Basin, notably in Cameroon, have been withdrawn, and the auditors who awarded them have lost their FSC licenses.

Principle 3 is controversial not because it is poorly conceived but because it is not a technical process. Crucially, it requires bringing two very different worlds together on relatively equal terms. Indigenous people should be able to participate in all the key management decision making concerning their traditional areas in spite of huge political, economic, and cultural differences. All very fine in theory, but how to achieve this in practice in relationships between discriminated, nonliterate, nomadic people and university-trained technocrats and businessmen?

A huge logging company operating in the area in which I did my doctoral research was one of the early companies seeking to obtain an FSC certificate.[4] While they had extensive expertise in the technical aspects of forestry—in how to reduce the impact of felling and road building, or how to calculate what was likely to be a sustainable harvest of commercial species, and so on—they did not have social expertise. In particular, they did not know what to do for the hunter-gatherers; they were technicians, not anthropologists.

To my surprise, they were suddenly interested in what I might be able to tell them so as to address the social principles of the FSC certificate they sought. Although they had been in the region for thirty years, it was difficult for them to understand the needs of the largely invisible hunter-gatherers. They did not know how or what to ask. This was a cultural and linguistic problem. Just as the hunter-gatherers find it almost impossible to communicate as a group with powerful others, so large, hierarchically organized institutions find it frustratingly challenging to communicate with the forest people.

It is enormously time-consuming and physically difficult to meet all the local groups associated with a particular territory. They live in small groups of twelve to sixty persons, often in remote forest dispersed over huge areas. Their locations are difficult to predict, as camps regularly move, sometimes simply to visit friends, sometimes as part of regular seasonal movements. The loggers' expectation of finding a leader then becomes the temptation to impose a "leader."

## Egalitarian Societies

As is well known in the anthropological world but rarely appreciated elsewhere, these are egalitarian people (Woodburn 1982). There are no leadership roles. No one has the right or authority to tell others what to do—even between parents and children. Here equality exists between men and women. Personal autonomy is so highly valued that it is considered rude to expect someone to do something for you, to boast, to criticize, or even to ask questions.

They are economically egalitarian. Production is consumed by all present, even those who never contribute, on the day that it is produced, and any excess production is shared on demand to anyone who asks for it (Woodburn 1982, 2005; Lewis 2005). Such social systems are extremely ancient to humanity and only effective when leading a hunter-gatherer lifestyle in an environment not subject to extreme seasonality. They have existed for hundreds of millennia in contrast to stratified farming or industrial societies. Despite this longevity, their rareness today has led to most people being ignorant of their organization, and often incredulous that such societies could exist at all.

Most people from hierarchically organized societies have great difficulty understanding that societies can exist without leaders or positions of authority. Almost universally, even in the international indigenous rights movement, outsiders expect or even demand to see the "leader" when interacting with such hunter-gatherers. Many Pygmy groups have found ways to address this by designating a spokesperson—*kombeti* among Yaka, *kobo* among Baka—to speak to outsiders. But to outsiders' surprise, these individuals can be either men or women and of highly variable age. Their skill is to speak well, and their authority goes only as far as their persuasiveness. This is characteristic of egalitarian societies the world over. Similarities have been well documented among the Batek in Malaysia (Endicott and Endicott 2007) and elsewhere.

If such an individual remains a spokesperson who is able to say what the majority thinks, the position is tenable. However, outsiders such as loggers expect such individuals to be able to negotiate with them and make decisions on behalf of the group in indoor urban-based meetings. At such occasions, substantial pressure is often put on spokespersons to do so. Then the "leader" is expected to return to his or her group and authoritatively enforce the agreements made with the outsider. Only much later is it realized that the decisions made have no legitimacy within the group. Most often such spokespersons do not even mention that they made any agreements when they return. It would be considered a great arrogance by their fellows. Negotiators from the outsiders often become angry with the "leader" and frustrated with the hunter-gatherers for being so "uncooperative."

This often results in increasing contact with the spokesperson as the negotiators seek to "build capacity." A major activity of the "social team" in many logging companies is to set up community associations to formalize the spokesperson's position. This is problematic, as it tends to increase the community's perception that the spokesperson is accruing benefits from his or her new status—always being visited by important outsiders, receiving gifts, and so on. This correspondingly reduces the spokesperson's legitimacy among his or her peers. If the spokesperson is suspected of receiving benefits without sharing them systematically with the rest of the group, he or she is quickly rejected and in extreme cases will be ostracized.

## Developing Solutions Collaboratively

I wished to avoid these predictable problems (described in Lewis 2001). Knowing the local language and which individuals were widely respected, I went to speak to them in their communities so that all present could join in. I asked what upset them about the way the loggers did their work. Men's and women's groups discussed sensitively and thoughtfully. Then, in turn, they explained to me that they did not mind sharing the forest with others so long as the loggers share properly (generally perceived as getting jobs to have access to the money that loggers mysteriously generate) and do not damage things that matter to them in the forest.

I asked what these things were. "We don't like workers camps in our hunting areas. We don't like it when bulldozers drive over the tombs of our relatives, or when they mess up springs that give us nice clean water. We don't want them to cut down our medicine trees, damage our sacred trees, or walk in our sacred places. We want our caterpillar trees kept standing so we can enjoy the caterpillars during the hunger season." And so on. I compiled a list and discussed the issues with other communities. Then I went back to the loggers.

Most of the resources that the Yaka wanted protected were unproblematic for the loggers. But one was especially controversial: caterpillars. Although not obvious to outsiders, the host trees for the most important caterpillars eaten by communities throughout the forest are large emergent sapelli.[5] Sapelli are also one of the most commercially sought-after trees. Large sapelli provide delicious, protein-rich caterpillars during what would otherwise be a hunger season when hunting is difficult because the first rains disperse animals widely in the forest. These caterpillars are a local delicacy and fetch a high price in local markets. Since women and the elderly are the principal collectors of caterpillars, they represent an important seasonal source of income to some of the poorest forest people. Additionally sapelli provide a range of important medicines effective against a number of common ailments (fever, pain, swelling, etc).

Clearly this was a major area of potential conflict with the loggers. How many trees would they insist on protecting? Would this prevent the loggers from making a profit? The possibilities were considerable. I had long discussions with the director. When I put a monetary value on the caterpillars—about 100 euros per sack in northern Congo, and a large sapelli can produce about five sacks a year—he suggested it would be easier for him to go into the caterpillar business than to continue logging.

I explained that the Yaka do not collect caterpillars from every sapelli, only from the ones located near places where they like to camp. Additionally it is the largest "emergent" trees that provide most abundantly, as they poke through the canopy and attract butterflies from far and wide, rather than the medium to large trees most popular with loggers. The issue provoked much discussion with the management group, but eventually all agreed that in the context of the FSC obligation to identify all potential conflicts of interest, they needed to map the trees that communities wanted to keep. If they were also trees the company wished to log, then they could negotiate with the community to resolve the conflict.

**Starting to Countermap**

The first stage was to identify who could be consulted to find out what resources the Yaka used in any given area of forest. I knew that each area of forest was the responsibility of a particular clan, so I assembled some key elders from the area of forest about to be logged and invited them into the map room of the logging company's offices. As I named and described the key indicators that the Yaka use to divide areas of forest—the turns of a particular stream or river, the position of salt licks, marshes, or lakes—they began naming which clan was responsible for each. As we slowly progressed from one area of firm land to the next, a new map of the forest began to emerge. Those large swathes of green were broken down into smaller areas bounded by the natural features of the forest, and each was named and had a clan responsible for it. For the first time, the presence of Pygmies became visible. It was now clear to all that they were present throughout the forest.

In a few hours, we had provided the loggers with a map identifying exactly whom they should consult before beginning to work in any part of the area about to be exploited. This is the key to ensuring that they had good guides who would be able to identify all the key resources in a given area. It also forms the basis for effective and efficient decision making during any possible future negotiations, dispute resolutions, or discussions of management issues.

Together with the loggers, we designed a procedure for doing the mapping that ensured that at least two men and two women, preferably accompanied by their whole families, would always lead the mapping teams.

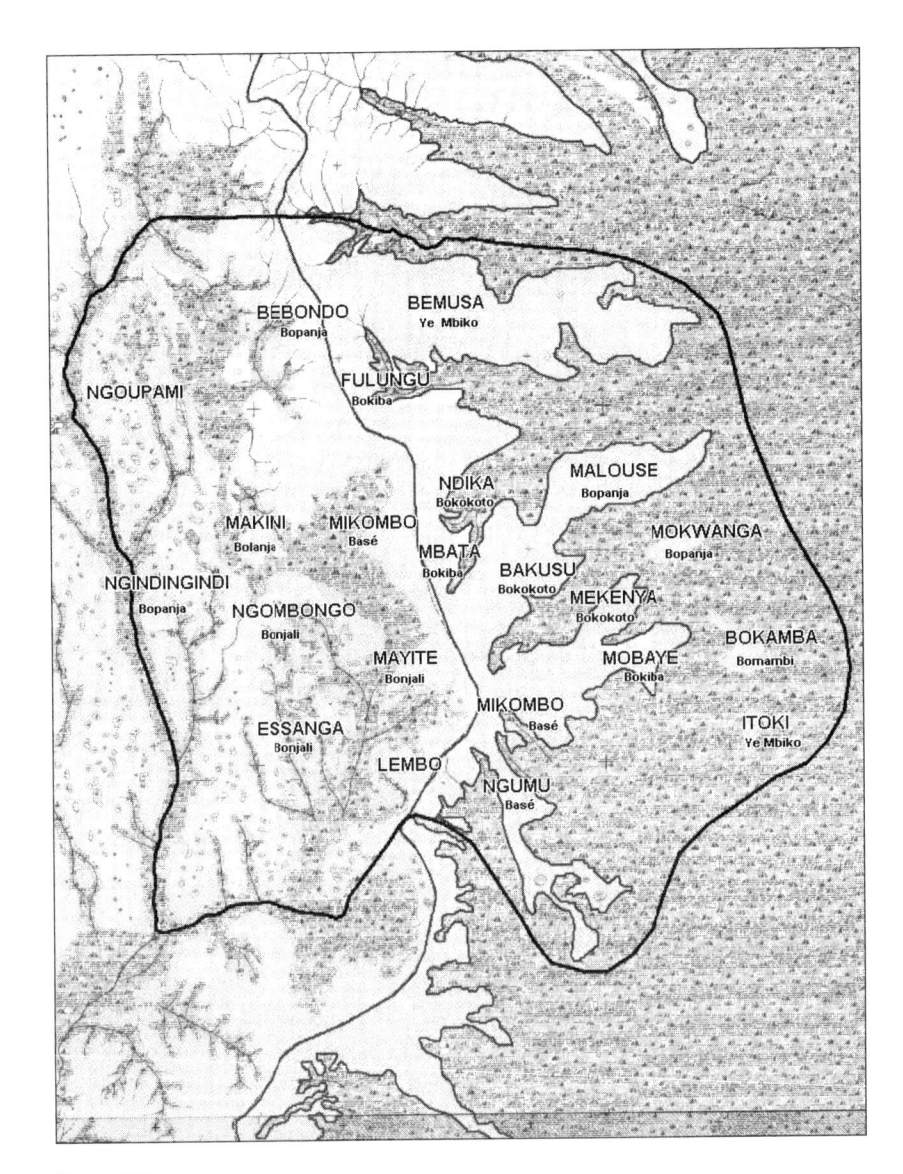

**Figure 7.3**
Part of the first map showing Yaka clans' forest areas.

Representatives of the logging company would accompany them to manipulate the GPS gadgets and take notes. This involved one person with the GPS collecting coordinates and assigning them way-points while another, sometimes the same person, wrote down in a notebook what each way-point represented: sacred tree, tomb, fishing area, or whatever the Yaka guides told them.

Back in the management office, these notebooks were transcribed, with the name of the researcher, date, way-point number and coordinates, resource indicated, and any other observations made, into Excel spreadsheets. These Excel files were then used to convert the coordinates into a format suitable for the geographic information system the management unit used to plan and organize forest activities. This labor-intensive process was prone to human error, confusing way-point numbers or coordinates, and required rigorous double checking, including sequentially mapping the way-points onto a map to check for consistency. Advances using this method were slow and therefore costly. Completing even relatively small areas took several months.

Once all were satisfied, the maps were printed out and taken back to communities for validation. Not having done more than indicate their resources to the men holding GPS machines, and the maps being composed of flat colors and labeled in writing, Yaka could not read them. Additionally, those showing them the maps did not try to explain them in terms that could be understood. As the first step to developing comanagement partnerships, there remained much to be done.

The system was inefficient, costly, error prone, and not understood by the forest people involved. A sustainable solution needed to collect precise quantitative data on resources located in inaccessible areas cheaply and exhaustively, using a method that would enable nonliterate people to understand what they were doing. The system had to be scalable over an area of forest half the size of Belgium. It seemed obvious to me that if local people collected the data, that process would make it cheaper and would allow them to understand the purpose of what they were doing and give them ownership of the data. But their innumeracy meant it was not possible for them to use a standard GPS.

## Developing Iconic Software

Remembering how Ingrid, my wife, had developed pictograms to label medicine bottles for use by the Yaka healers whom she trained to run a mobile pharmacy, I wondered if it would be possible to do a similar thing for a GPS gadget. The Forest Trust (a UK charity), which was supporting this work,[6] put me in touch with a UK software company called Helveta that specialized in chain-of-custody tracking for products used by industrial companies seeking legal certificates of origin or forestry certification. I talked to the director about my idea for iconic software loaded onto the ruggedized handheld computers they already used. To my surprise and great pleasure, he agreed to make up the software if I designed it.

Using our knowledge of how Yaka divide up activities and what they had told me about the resources they were concerned about, Ingrid and I developed icons to represent these different activities and organized them as Yaka would. So the front screen depicted a man with a spear (hunting areas), a couple with ax and basket (gathered resources), a group of people (social and religious resources), a fish (fishing places), and a farm. Pressing on one of these icons led to a new screen with icons depicting characteristic actions associated with extracting a particular resource (a boy drinking springwater from a leaf, for instance, or a woman skewering caterpillars on a stick for smoking). Our years living in the forest meant that we were sure the icons would be obvious to the Yaka. If the software was to work, it had to have an intuitive logic for forest people.

The handheld computer onto which the software was to be loaded was developed for military use, so it was extremely robust. It could be dropped onto concrete, jumped on, submerged three meters underwater, and even burned while still keeping the data collected safe on its hard drive. It used a quick and extremely accurate GPS and was only slightly larger than the GPS gadgets normally used under the forest canopy. This made it ideal for use by the tough Yaka as they walked their humid forest.

Within ten minutes, almost anyone could start using the handheld device, and within half an hour, a sensible person would be fully competent. The only difficulties encountered were among some older people with poor eyesight. They had difficulty seeing the icons on the small screen. Younger people would quickly come to their help, showing them what to press. During mapping with such elders, the young people would manipulate the handheld device and so learned from the elders about resources and areas they may not have previously known.

Importantly, the new system overcame a number of key problems associated with human error and efficiency. Data were automatically registered with location and type of resource on the hard drive and could be transferred by Bluetooth or cable to a suitably configured laptop, even deep in the forest, in less than a minute without any mistakes. First visualizations were on Google Earth; later better-quality satellite images were used. These proved

**Screen 1**         **Screen 2**         **Screen 3**

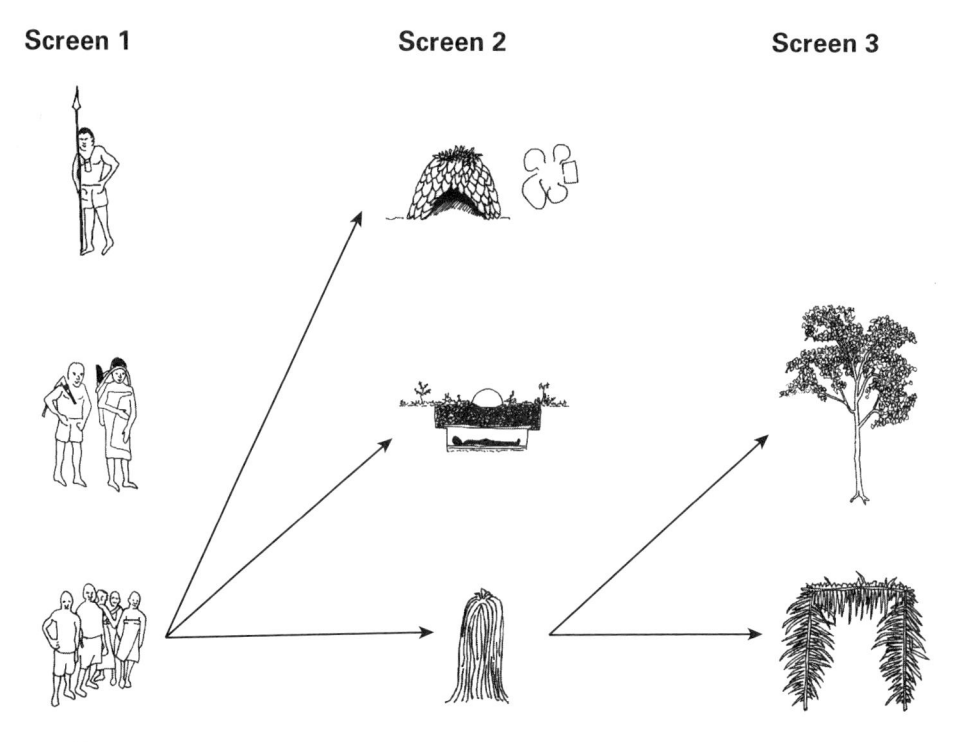

**Figure 7.4**
Examples of icons used to map a sacred tree or sacred area.

much easier for people to understand than the abstract flat colors symbolizing different vegetation types or geographic features on previous maps.

Now the quality of the mapping could be checked as mapping progressed. If all the points were on a straight line, then clearly people were not properly covering the territory but simply walking along a path and pressing icons from time to time. The maps could be discussed with the community mappers, and elements could be questioned: why is nothing recorded here, why so many resources here, and so on? In this way, participants could better understand the point of what they were doing and become increasingly confident at interpreting the maps by understanding more of the process by which the resources they were identifying were turned into pictures on a piece of paper. Finally, we had a tool that would allow anyone, regardless of literacy or language, to map the forest people's key resources.

In three months, every cutting block for the following year had been mapped by local inhabitants and their key resources accurately and clearly recorded. Previously this would have taken nine or more months to complete. Since 2006 the system has been integrated into the procedures that the logging company follows to prepare a cutting block and ensure that local peoples' resources are not damaged during logging. The system continues to work effectively, and the handhelds have not broken or crashed.

## Maps Do the Talking

Rather than Yaka hunter-gatherers being taken into the intimidating atmosphere of the management offices, the maps would go there for them. Company staff would translate the icon maps into ArcView files that were compatible with the GIS software they used to organize forest exploitation. They could compare the resources that Yaka wanted to protect with those they had planned to cut down. Wherever there was a conjunction of a resource they wanted to cut and one the people wanted to protect, the resource was taken out of the cutting schedule. In theory, if the company felt that they had to remove a resource that forest people wanted to keep, they could begin negotiations at this stage. In fact, this has not been necessary.

Even in areas that are very rich in key resources for Yaka it was possible to log without damaging these resources. The company's head of management told me that the sapelli trees that Yaka wanted kept for caterpillars were often too big and so likely to be rotten inside, and too difficult to get out of the forest for the loggers to be interested in them. Those that were not could be left out, since they had a 15 percent margin for choosing which trees to submit for the annual permit to the Ministry of Forests from the total number of trees initially identified for felling. Trees are withdrawn from cutting schedules for a variety of reasons—split trunks, rot, or growing on steep inclines, and now also because forest people need them. So despite the large number of trees in this cutting block, CIB was still able to exploit efficiently and profitably without damaging Yaka resources.

Once the management team had identified all the resources that required protection from the company's activities, they produced these maps to inform forestry workers of what they should be careful to avoid during exploitation. These maps left off much information to protect other resources and knowledge about good hunting areas, fishing sites, and so on from workers. Meanwhile management representatives went back to the community with pots of pink paint to clearly mark every area or tree to be protected.

By using the icon-driven software and producing their own maps, non-literate Yaka have obtained an advocacy tool that communicates their main concerns to CIB's top management more efficiently than anything they

146  Jerome Lewis

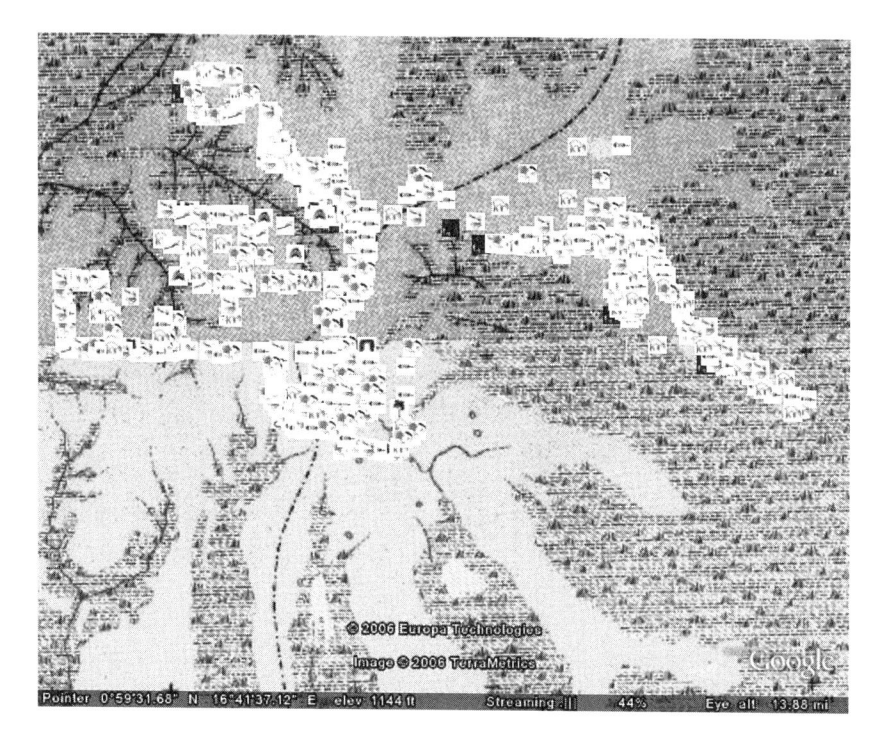

**CARTOGRAPHIE SOCIALE ZONE IBAMBA UFA PKL ET TKLK**

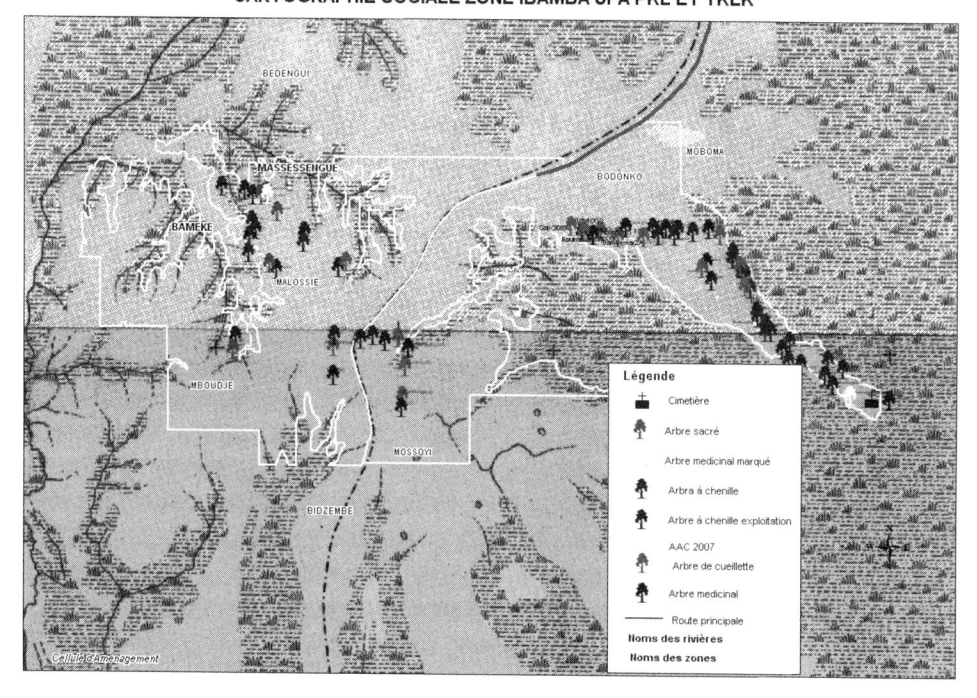

Légende
- Cimetière
- Arbre sacré
- Arbre medicinal marqué
- Arbre à chenille
- Arbre à chenille exploitation
- AAC 2007
- Arbre de cueillette
- Arbre medicinal
- Route principale
- Noms des rivières
- Noms des zones

# Making the Invisible Visible 147

thought possible before. They are pleased that this tool enables them to be absent when communicating their concerns, since they feel so uncomfortable during meetings in towns. Where previously they were invisible and ignored, now their presence in the forest and their dependence on its resources are recognized and respected. This is a huge step forward.

Much remains to be done. Only a handful of logging companies are FSC certified in the Congo Basin, and nowhere in the tropics has the promise of FSC been fully achieved. The problems are many and complex, from corrupt institutions, illegal logging, and auditors with different standards for judging compliance, to companies continuing to ignore indigenous peoples in nonexistent or poorly designed or implemented social programs. However, participatory mapping with indigenous peoples has become standard practice among logging companies seeking certification across the Congo Basin. In effect, the rights of indigenous peoples to their land and resources are gaining de facto acceptance despite the state's dismissal of them.

The Yaka, like so many people in the Congo Basin, are facing enormous change. Conservationists exclude them from rich forest areas at the same time as logging roads open up remaining forest areas to outsiders, and global warming is changing rainfall patterns and the seasons that were previously such a reliable guide to the appearance of wild foods.

To facilitate forest people taking stock of this new situation, together with the Forest Trust, we have launched a community radio station broadcasting uniquely in local languages:[7] Biso na Biso, literally "between ourselves." Biso na Biso is facilitating communication between numerous small camps distributed across the forest so that they can develop their own understanding of the situations facing them and share news, concerns, insights, observations, solutions, and analyses. Eventually we hope to create an international network of community media outlets in tropical forests of the world so that forest people will increasingly be in a position to negotiate confidently with outsiders in an informed way to secure their long-term interests.

## The Eyes and Ears of the Forest

The success of the mapping project in Congo spawned other projects in different parts of West and Central Africa. In Central African Republic and

**Figures 7.5 and 7.6 (opposite page)**
One of the first Yaka resource maps with icons next to the one of the same area for logging workers. Courtesy of CIB cellule d'amenagement.

Cameroon, mapping has been used to show that hunter-gatherers require access to their customary forest recently encompassed by protected areas, in Nigeria to track rare primates and other fauna in a national park, and in Cameroon to monitor logging.

The Cameroon project reveals some of the directions of change, and though I can describe it only briefly here, Lewis 2007 provides more detail. The forests of Cameroon are subject to extensive legal and illegal logging. Many communities, including Baka Pygmies, are losing important trees, such as moabi[8] and sapelli, which they depend on for fruit, caterpillars, medicines, and oils, and for the modest incomes generated by selling these products in local markets. In addition to industrial loggers, artisan loggers make incursions into community forest, where they may destroy or damage vital resources used by local people.[9] Until now, local communities have had little support in facing up to these serious threats to their livelihoods. They, like the Cameroonian government, the European Union, and others, want these practices to stop.

As part of an effort to improve forest governance, the Cameroonian and British governments supported a redeployment of the system developed in Congo. The project's objective was to enable local forest people to monitor logging activities in their forest areas regardless of their background, education, or language and so contribute to an independent means of verifying the enforcement of forestry law.[10] Locally collected data on logging activities are periodically uploaded to a secure Web site via satellite link. Project partners, including Cameroonian forest law enforcement agencies and local NGO partners, can access the site to gain up-to-date information to monitor and control logging activities. In addition to increasing the government's forest-monitoring capacity, the Web site takes a first step in building a dialogue between government, NGOs, and communities about forest management. It also provides an accessible platform to audit and demonstrate governmental commitment to good governance.

In southeastern Cameroon, I did not have the same intimacy with local practices in the forest as I did in Congo, so I could not rely on my own knowledge to develop appropriate icons. So that the software would be as intuitive and self-evident to the new users as it was for Yaka in Congo, I configured a prototype iconic decision tree and took it out to forest communities for testing, accompanied by a software engineer from Helveta and local NGO partners. The new configuration included most of what had been used in Congo but added a section for monitoring logging.

Once encamped in the first small village on our circuit in southeastern Cameroon, we presented the icons one by one to the assembled villagers.

# Making the Invisible Visible 149

Rather than telling them what each icon meant, we asked them to tell us. Performing this test in several different ethnic areas allowed us to check that the icons were intuitively obvious across language groups. During each session, we noted problems to be modified later. Once people were clear about all aspects of the software, including how and when to press the different icons on the screen, we formed three working groups to take the handhelds into the forest around the village. Depending on the ethnic mix of the village, we would compose groups to reflect this. There was always one group of women.

Each group would be accompanied by members of the team during a two- to three-hour walk as they visited resources in the forest around the village and mapped them using the handhelds. Each member of the four- to five-person working party would be encouraged to use the handheld to provide as much feedback as possible on the icons and decision tree. As people tested the software, they pointed out problems: things they wanted that were not there, icons that were unclear or could be improved, and so on. All of this was noted. When we returned to the village, the software engineer Simon Bates and I, surrounded by local commentators, would spend a few hours drawing new icons and updating the software to reflect the comments and changes desired by the community. This enabled us to test the new icons and decision tree the following day in the next community.

This method of participative software development was extremely successful. As a result of this iterative development, the icons for monitoring logging became more and more tuned to local realities. A tree stump on the front page led to further screens that allowed the user to geo-reference felled trees, abandoned logs, and trees marked for future felling, and to record the diameter of the tree or log using color-coded lengths of string and even indicate if the logging was conducted by an industrial logger (bulldozer tracks) or an artisan logger (abandoned planks). By the third community visit, people no longer requested changes. We were now confident that the software could be deployed among users speaking a range of different languages.

## Where Next?

As new problems are identified, appropriate software builds can provide local people with the means to monitor the issue. To facilitate this kind of redeployment, in the future we hope to develop an authoring tool that easily enables a user to customize pictogram decision trees or create new ones depending on the cultural and environmental context and issues to address.

In the next stage of work with the Yaka in Congo-Brazzaville, the handhelds already being used for resource mapping by the logging company will be reconfigured to address another problem. Yaka hunters are extremely concerned about overhunting by commercial poachers. This is also a major preoccupation of conservationists and the logging company, but they have never found an effective means of using the Yaka's knowledge of poachers' whereabouts to control them more effectively. In January 2010, we plan to build a new set of software specifically for recording evidence of poachers' activities. Small teams of Yaka hunters will visit forest areas and record data on poaching activities. At the management office, the data will be downloaded and sent by e-mail to the conservationists managing the ecoguards. Patrols will be organized to follow up on the information gathered. This project is funded by native North Americans (First Peoples Worldwide) who are supporting other indigenous peoples around the world to regain a meaningful role in managing their environments.

We have two more projects, both currently seeking funding. One seeks to empower forest people to become key players in the emerging environmental services and carbon-trading markets by enabling them to provide high-quality data monitoring the condition of standing carbon stocks (i.e., trees), water levels, and variations in animal populations and other environmental services provided by dense tropical forests. Unless local forest people have some opportunity to play a role in these emerging economies, they will simply be bypassed, and yet more of what they consider to be their heritage and birthright will be denied them, and the benefits enjoyed by others elsewhere.

The second project seeks to better use the potential that clear visualization of mapped data has to help people make well-informed decisions. This is particularly important for future forest management in the context of climate change and rapid human-induced environmental change. Rather than simply provide static maps, we plan to create interactive, change-displaying, easy-to-use, portable digital maps based on the icons that people use to collect the data. The maps would be capable of being used to analyze or measure change quantitatively and for producing reports with data represented in the form of maps, tables, or charts appropriate for the needs of relevant authorities making decisions and taking action.

Although planned to be GIS accessible to nonliterate users, on the one hand, the resulting maps will try to organize the environment in ways defined by the people. Thus we envision that they will be able to store multimedia data at single geographic points to show changes over time, or to store large files containing myths or songs relating to the landscape, and so on.

The aim is to create tools that enable forest people to monitor their environment in ways that are useful to many different interested parties, and in so doing make it more likely that forest peoples' concerns are taken into account by national and international decision makers. By using models based on local ways of categorizing the world, these tools will also facilitate a broader understanding by forest people of the changes rapidly sweeping their forest regions. By visualizing the impact of changes in their traditional areas in new ways, forest people will educate themselves, and us, in how their ecosystem is changing, and hopefully stimulate us all to think about why and what to do about it.

## Notes

1. Silent trade is characterized by the parties not meeting during the transaction. Rather, goods are deposited in secret and left for one party to collect. The receiving party then leaves goods in the same place for the instigators to collect at a later time.

2. While this has been changing rapidly in recent years as touch screens and symbolic codes are increasingly developed, this was not widespread just a few years ago.

3. Slightly larger than half the size of Belgium.

4. Congolaise Industrielle de Bois (CIB), with concessions covering 1.3 million hectares.

5. *Entandrophragma cylindericum*.

6. With a World Bank Development Marketplace grant.

7. Funded by the World Bank Development Marketplace and the Chirac Foundation, and supported by Open Air Radio at the School of Oriental and African Studies, University of London.

8. *Baillonella toxisperma*.

9. Artisan loggers are groups of men with chain saws and frames that they use to cut planks directly where the tree falls. The planks are then carried individually out of the forest to roadsides for transport and sale.

10. This project was funded by the British Commonwealth and Foreign Office.

## References

Endicott, Kirk, and Karen Endicott. 2007. *The Headman Was a Woman*. Long Grove, IL: Waveland Press.

Lewis, Jerome. 2001. Forest people or village people: Whose voice will be heard? In *Africa's Indigenous Peoples: "First Peoples" or "Marginalized Minorities"?*, ed. Alan Barnard and Justin Kenrick, 61–78. Edinburgh: CAS.

Lewis, Jerome. 2002. Forest hunter-gatherers and their world: A study of the Mbendjele Yaka Pygmies and their secular and religious activities and representations. Ph.D. thesis, University of London.

Lewis, Jerome. 2005. Whose forest is it anyway? Mbendjele Yaka Pygmies, the Ndoki Forest, and the wider world. In *Property and Equality*, vol. 2: *Encapsulation, Commercialisation, Discrimination*, ed. Thomas Widlok and Wolde Tadesse, 56–78. Oxford: Berghahn.

Lewis, Jerome. 2007. Enabling forest people to map their resources and monitor illegal logging in Cameroon. *Before Farming: The Archaeology and Anthropology of Hunter-Gatherers* 2. http://www.waspress.co.uk/journals/beforefarming/journal_20072/news/2007_2_03.pdf.

Lewis, Jerome. 2008a. Ekila: Blood, bodies, and egalitarian societies. *Journal of the Royal Anthropological Institute* 14:297–315.

Lewis, Jerome. 2008b. Maintaining abundance, not chasing scarcity: The big challenge for the twenty-first century. *Radical Anthropology Group Journal* 2. http://www.radicalanthropologygroup.org/new/Journal_files/journal_02.pdf.

N'zobo, Roch Euloge, Roger Bouka Owoko, and Alain Oyandzi. 2004. *The Situation of Pygmies in the Republic of Congo*. Rainforest Foundation and OCDH. http://www.rainforestfoundationuk.org/National_analysis_of_the_rights_of_indigenous_people_in_the_Republic_of_Congo.

Schadeberg, Thilo. 1999. Batwa: The Bantu name for the Invisible People. In *Central African Hunter-Gatherers in a Multidisciplinary Perspective: Challenging Elusiveness*, ed. K. Biesbrouck, S. Elders, and G. Rossel, 21–40. Leiden: CNWS, Universiteit Leiden.

Woodburn, James. 1982. Egalitarian societies. *Man: The Journal of the Royal Anthropological Institute* 17:431–451.

Woodburn, James. 2005. Egalitarian societies revisited. In *Property and Equality*, vol. 1: *Ritualisation, Sharing, Egalitarianism*, ed. Thomas Widlok and Wolde Tadesse. Oxford: Berghahn.

# 8 Assembling Diverse Knowledges: Trails and Storied Spaces in Time

David Turnbull and Wade Chambers

**Part I: Theoretical Overview**

At least since Plato, critical thinkers have been aware of a key problem at the heart of Western civilization. Although knowledge is revered and given centrality as the keystone of all that is good and true, there is no agreement on how to define it, and knowledge comes in many guises. Michael Polanyi, for example, famously pointed out that "we know more than we can tell." He called this kind of knowledge "tacit knowledge"—the kind of acquired, taken-for-granted skills that are essential to using computers, for example (Polanyi 1958). But precisely because tacit knowledge is acquired almost unconsciously and can only be expressed with careful introspection, it makes it hard for the younger generation to teach the older generation how to send an e-mail or get on the Internet. At the same time, those very computers and their interconnections through the World Wide Web have vastly augmented our capacity to assemble and access knowledge.

Indeed, we live in an age where the ever-increasing volume, power, and significance of knowledge have led to a declaration of a "knowledge economy." In this era of "cognitive capitalism," the most seductive dream is of a knowledge panopticon in which all knowledge can be assembled in one universal archive. But this dream is proving hard to realize; the torrential flow of new knowledge and its unremitting commercialization make knowledge extremely difficult to manage, store, redistribute, and use, and even more acutely difficult to define, monitor, and evaluate. Moreover, the connectivity of the Internet is also creating the possibility of new forms of knowledge, some of which we are only just beginning to discern.

The modern condition, then, is one of increasingly fragmented and sequestered pools of knowing and unknowing, with wisdom and a sense of the totality in short supply. Recognizing these difficulties has led many to suggest that the way out of the morass is through a "knowledge commons," to reestablish knowledge sharing as a complex ecosystem (Hesse and Ostrom 2007).[1] But the situation is now much more difficult to resolve than simply dealing with the difficulties of knowledge proliferation and evaluation.

While the problem of knowledge has always lain half-acknowledged at the heart of both Western epistemology and its political economy, it has largely been assumed that scientific knowledge at least could become unified. Now we must accept the possibility that science not only is ununified but may be ununifiable (Galison and Stump 1996).

Furthermore, that disunity is radically compounded as we start to recognize the value and importance of previously dismissed indigenous knowledges (Turnbull 2000). Finally, we are beginning to realize that the knowledge-making process, even in the West, has always been locally contingent in character, rather than universal (Chambers and Gillespie 2000). Thus the "problem of knowledge" becomes the "problem of multiple and diverse knowledges." The approach of this paper is to reimagine a knowledge commons, using technologies of computing, that would allow for the inclusion of diverse indigenous knowledge systems from cultures around the world.

Working with this multiplicity requires new technologies that enable the storage, maintenance, retrieval, and comparison of any knowledge-bearing text, object, or performance, indeed, any entity with narrative, semiotic, or tacit components. In practical terms, this means that researchers, scholars, curators, and teachers must be prepared to manage and interpret knowledge that is embodied not only in the standard text but also in stories, dance, calendars, maps, architecture, textiles, hunting and farming practices, weapons, rock art, trails, spaces, stone placements, or any other form in which knowledge may be fixed, discerned, or performed (Chambers 2008).

One important dimension of this approach is the assumption that indigenous knowledge traditions cannot be subsumed in specific language texts, whether written or spoken; or in single objects, such as maps, totem poles, and quipus; or in environments, whether natural, altered, or constructed; or in performances, such as agricultural practice, ceremony, and dance. Rather, knowledge is only fully apprehended by exploring connection, context, relationship, kinship, and reciprocity. In indigenous culture, all knowledge is ecological—a multiplicity of narratives, local practices, artifacts, and ontologies taken together with their intersections and interactions, physical and cultural. In part 2, "The Shape of the Indigenous Knowledge Commons," we describe our attempts to design new teaching and research spaces for the Internet, based on the hypertext multimedia resources of the World Wide Web. The initiative arose in direct response to the felt need of teachers, scholars, and curators who face the problem of bringing disparate knowledges together.

A fundamental reason why it is of paramount importance to deal with indigenous knowledges is that in the last century six hundred indigenous languages vanished forever. Their eradication continues at the rate of one language every two weeks. By the end of this century, 90 percent of indigenous languages will have ceased to exist (United Nations Permanent Forum on Indigenous Issues 2008). With linguistic diversity goes cultural diversity, with cultural diversity goes biological diversity, and at the heart of this biocultural complex lie indigenous knowledges (IKs). When languages die, knowledge and, equally importantly, practices go with them (Turnbull 2009; Harrison 2007; Evans 2009; Maffi 2005). The future of IKs are thus on a knife's edge.

The recognition, understanding, and social-commercial realities of IKs are, however, undergoing profound changes. In India the government has recognized the commercial value of traditional medical knowledge and established a massive database to document classical herbal formulations and their therapeutic use. The Traditional Knowledge Digital Library (TKDL) aims to establish traditional medical knowledge as "prior art" and thereby resist biopirates patenting indigenous knowledges, as was successfully done in the groundbreaking case of turmeric.[2] At the same time, a systematic and critical evaluation of rural and local health traditions, the Foundation for the Revitalisation of Local Health Traditions (FRLHT), has been initiated along with the opening of the Centre for Indian Knowledge Systems in Chennai (CIKS).[3]

In 2004 the South African government announced an "African renaissance" in which IK would be the platform for sustainable economic development not only of South Africa but of the African continent as a whole. A division was created within the Department of Science and Technology devoted to affirming, developing, promoting, and protecting IK (Green 2008, 2009).[4]

While the recognition of the importance of IKs has risen around the world, the Americas have seen the most profound transformation in IKs' reconceptualization. A real grassroots "indigenous resurgence" is occurring across all social, political, and cultural arenas. It can be seen in the rise to power of national governments with strong indigenous representation in Venezuela, Brazil, Ecuador, and Chile, but especially Bolivia (Boccara 2006; *Economist* 2004). In North America, observers have described a "renaissance" in native arts, native scholarship, and the way the native community speaks to power. Some of the more important but less obvious dimensions of this resurgence have directly and indirectly focused on recognizing and conceptualizing IKs.

In the popular imagination, South America in general and the Amazon in particular are still represented as the "New" World, the last frontier where wild savages live in pristine nature. It is not unusual to read headlines like "Isolated Tribe Spotted in Brazil"[5] or "The Amazon Tribe That Speaks Language without Numbers."[6] Though the colonial mythology of the "new," the "pristine," and the "primitive" has long since been debunked in academia, what has yet to be absorbed into full consciousness is the revolutionary impact of recent work in historical ecology and philosophy that puts South American IKs in a very different light.

Enrique Dussel, the Argentinean Mexican liberation philosopher, established a key philosophical move that allows for the conceptualization of American Indian knowledges in their own terms, rather than being subordinated in Western modernist categories. Dussel gives modernity its European originary moment as 1492, when Europe defined itself as the center of world history and the arbiter of all knowledge in contradistinction to, and negation of, the New World other. Dussel proposes the creation of a "transmodern" space where differing incommensurable knowledge traditions can work together to create new knowledge, a space that allows the dynamic of "agonistic pluralism" "in which modernity and its negated alterity would co-realize themselves in a process of mutual, creative, fertilisation" (Dussel 1993, 1996, 2009; Turnbull 2003, 227–228). Other South American writers have also developed a similar position, which has been described as the "dialogical pluralism of interculturality" (Ribeiro and Escobar 2006, 4; de la Cadena 2006, 217–218; Bowker 2000a).[7]

What makes differing knowledge traditions important but problematically incommensurable is their differing ontologies and epistemologies. Viveiros de Castro, a Brazilian anthropologist, has articulated a relational epistemology, "Amazonian perspectivism," that stands in opposition to the classical ontological position of modernity in which nature and culture, humans and the environment, us and them, are divided along apparently naturalistic lines. From an indigenous American perspective, humans and animals are not separate but in a common category having a spiritual unity and sharing many "natural" characteristics; where they differ is in their perspectives on the world. In this ontology, the point of view creates the subject, resulting in "multinaturalism," or multiple natures, rather than multiculturalism (Viveiros de Castro 1998, 2004). This raises difficulties for knowledge assemblage, because, as Bruno Latour has pointed out, modernity constitutes itself through constructing these great divides; allowing multiple ontologies would be to undo the divides and consequently to undo modernity (Blaser 2009; Povinelli 2001; Latour 1993, 2004).

However, modernity in its conception of itself is further undone by another South American revolutionary overthrow of the orthodox humanity–environment relationship. Looming large in the European imagination, along with pristine nature, was El Dorado, the city of gold, discovered by Francisco de Orellana in 1542. Orellana claimed to have encountered inhabited cities stretching for miles along the Negro and the Amazon rivers—cities that later explorers were completely unable to see, reporting nothing but dense jungle and savage hunter-gatherers. Thus the Europeans assumed that the reported population centers had never existed. This assumption confirmed several prejudices: that the soils and the peoples of the Amazon were simply too primitive to support anything approximating civilization, and civilization was an essentially Western development, dependent on agriculture and sedentism.

Recently such preconceptions have undergone a radical transformation. The Amazon rain forest, though still under threat of destruction and annihilation and still a vital component of the global climate system, is no longer the "last great wilderness," nature untouched by humans. It is now recognized as anthropogenic, a "cultural forest," a product of deliberate human activity and knowledge (Balée and Erickson 2006; Denevan 2001; Erickson 2003; Sandor et al. 2006; Mann 2000, 2005). Rather than the standard Western understanding in which nature and culture become divided as agriculture is invented and humans separate themselves from nature, historical ecology in the Amazon reveals the role of both history and human agency in a complex coproduction of the landscape.

Contemporary indigenous Amazonian cultures have comprehensive, detailed knowledge of local plants, animals, food production, and soil types but disclaim any knowledge of how *terra preta* is made (German 2003). What Erickson, Balée, Denevan, and others have shown is that the early inhabitants of the Amazon were able to transform the otherwise inadequate soil, through the addition of charcoal, pottery shards, manure, and fish remains, into a rich, self-reproducing dark earth—*terra preta*. This soil, along with massive earthworks, mounds, causeways, canals, and fish farms, allowed the development of extensive, complex, heterarchical societies. These societies, which were apparently devastated by European diseases, quickly became invisible and largely unknown, even to the remaining native inhabitants. Importantly, the question of dark earths, their origin and suitability for contemporary adoption for carbon storage, is an IK issue that now raises the problem of critical evaluation in the quest for climate change solutions. James Lovelock thinks it is humanity's one last chance (Vince 2009); by contrast, George Monbiot argues that biochar is madness.[8]

IKs, especially in the indigenous American case, fit fairly well within the constructivist-interactionist paradigm that has become the key alternative to structuralist-functionalist explanations. Examples of such explanations range from actor network theory, complex adaptive systems, and niche construction to developmental systems, but what they all share is the common recognition of a relational epistemology and, given their emphasis on the interaction of all the components of a network or system, the ubiquity of agency (Odling-Smee, Laland, and Feldman 2003; Oyama, Griffiths, and Gray 2001a). This radically undercuts both the dichotomizing of the great divides and problems of intentionality, determinism, and origins that beset many of the alternatives.

The key question is: how can these developments and insights be brought to bear on analyzing the effects of aligning information and communications technologies with indigenous knowledges while at the same time maintaining the possibility of critical comparison and evaluation? It is seductively easy to put IKs beyond criticism because they have special virtues: spirituality, attachment to land, a relational epistemology, a oneness with nature. Equally it is seductively easy to argue that Western science is inherently problematic because it has historically served colonization, exploitation, and the destruction of nature. Such seductions could lead one to fear the ICT-IK alignment as a threat to the preservation of traditional IK values. Alternatively, IKs picking up ICT might be seen as unproblematically virtuous, an example of hybrid resistance giving subordinated traditions a voice.

That IKs cannot be accepted uncritically is shown in the case of South Africa, where elevating IK approaches above Western science was accompanied by the former president Mbeki's denial of scientific understandings of the etiology and treatment of AIDS, thus increasing suffering and death among the infected. Equally, it is necessary to find firm ground on which to evaluate arguments of the opposite kind, which blame the knowledge of native people for their own downfall. For example, in his recent polemic *Collapse: How Societies Choose to Fail or Survive*, Jared Diamond suggests that in many cases, such as Easter Island, local IKs and cultures have been the cause of social collapse. Countering Diamond's claims, good scientific evidence indicates that the total elimination of the island's palm trees was caused not by the blind, unthinking erection of more and more sculptures but by the introduction of rats that ate the palm tree seeds (Diamond

2005).[9] In other words, both Western science and indigenous knowledge require critical evaluation in the contemporary world.

While it is obviously true that IKs have been driven close to extinction under the impact of Western knowledge and its institutions, many people now believe that one way of mitigating the effects of Western science on the environment is to adopt some indigenous understandings. Clearly it is necessary to address fundamental issues about knowledge in all traditions, both Western and non-Western. If we are to abandon the idea that knowledge is singular, monumental, and represented by the propositional statements of Western science, if knowledges are to be recognized as multiple, diverse, and hence in some sense profoundly different and incommensurable, how, then, are they to be authorized and validated, and by whom, and by what standards?

The International Council for Science's rejection of UNESCO's call to incorporate IKs, on the grounds that they cannot be assembled and evaluated using the methodologies of Western science (Bowker 2000b; Bates et al. 2009), creates a classic double bind. If we accept that Western modes of evaluation and assemblage do not or cannot be applied to traditional knowledges, then how can we avoid having to uncritically accept what may be superstition or oppressive beliefs like AIDS denial or the inherent inferiority of women (Nanda 2003)? On the other hand, if Western technologies like ICT and Western standards and categories of thought are applied, can the differing ontologies and epistemologies of IKs survive to provide the sought-after cultural diversity? If they are so homogenized by submitting to Western standardization and evaluation, is there any diversity left? Can IKs be understood in a way that enables them to be represented, assembled, and translated without loss, and on what grounds can the varieties of knowledge traditions be compared? If IKs are not uniformly valuable and virtuous, how is good to be distinguished from bad?

To resolve this difficult dilemma, we need to draw on recent thinking about complex adaptive systems and link it to narrative and movement in the construction of knowledge, place, and space, and the ways in which the distributional, the dialogical, and the diverse are key emergent processes in the creation of new knowledge.

### Complex Adaptive Systems

Much of the thinking around the idea of complex adaptive systems suggests that we must develop new ways of working with differing knowledge traditions, and we must learn how to create forms of knowledge assemblage and production or databases that, rather than having one unified, preset system of taxonomy and ontology, would instead permit multiple, diverse ontologies to interact. The reasons for examining complex adaptive systems are thus at least twofold: first, the common focus on multiplicity or diversity; second, the recognition that if no one ontology is to be privileged, then the outcome must of necessity be emergent.

But, of course, such databases also aim to provide ordered access to an assemblage of objects and materials in one time and place. Thus the project of working with disparate knowledge traditions is located in oppositions that are perhaps even more definitional of the age than those we mentioned earlier. Inherent in any attempt to conceive the nature of knowledge production as a dynamic process are oppositions that include unity and totality versus multiplicity and diversity; preset laws, rules, protocols, standards, methods, and algorithms set against emergent plans, methods, rules, and protocols; and an underlying reality that is predominantly physicochemical versus one that is biological and historical.

Systems theory, though it has a venerable history going at least as far back as Bertalanffy (1968),[10] is neither particularly well recognized nor a unified, coherent body of theory. It thus comes in a variety of guises, depending on the parent discipline, the central problematic, and the corresponding selection of high-level functional characterizations. It is variously discussed as the science of networks or network theory (Barabasi 2002), self-organization (Bak 1996), emergence (Johnson 2001), developmental systems theory (Oyama, Griffiths, and Gray 2001a), complexity theory (Lewin 1992; Waldrop 1992), complex adaptive systems (Holland 1995), small world theory (Watts 2003; Buchanan 2002), and chaos theory (Gleick 1987).

What the various theories have in common is some notion of complexity and of systems. They can be differentiated in a variety of ways; for example, some of them are concerned with emergence, others with self-organization. However, no theory makes an attempt to cope with the totality of the field of systems, nor is there any agreement on how to define complexity. Some commentators see this fragmentation and disunity as a sign of radical incoherence and an indicator of the field's inevitable collapse, or even as a walking corpse on the trail of faddish theories like chaos and catastrophe (Horgan 1995). But to philosophers of biology and science studies generally, the situation is symptomatic of science itself. Natural selection, for example, despite its ubiquity and embeddedness, has no fully articulated theoretical basis that is accepted across all the areas of its application. The very breadth and strength of natural selection as a theory means that it is "an indeterminately bounded process"; its various theoretical formulations reflect differing constellations of general high-level functional characteristics from Darwin through Dawkins to Oyama, on to Jablonka, Margulis, and Wimsatt (MacLaurin 2005).[11]

From our perspective, two discourses of complexity correspond to the oppositions mentioned earlier. There is one discourse of unity, of deep simplicity, which can explain not only all of nature's diversity, physical and biological, in terms of simple algorithms, dissipative structures, fractals, network dynamics, and power laws, but also all of culture's diversity (Gribbin 2004). There is another parallel discourse with the contrasting high-level functional characterization of plurality, in which multiplicity is the essence and dynamic at every level of complex systems (Goldenfeld and Kadanoff 1999). We now look briefly at the similarities and differences between these two discourses of systems theory to explore what they suggest for the project of working with incommensurability.

Systems theory is not always explicit about it, but in the end, or the beginning, it is, of necessity, an attempt to answer the really big metaphysical questions: Why is there order and structure in the universe? How do we account for the properties of things, along with the attendant epistemological and ontological questions? What sorts of things are there in the universe? How do we know them, and how do we know we know them? These questions are just as problematic for the social as they are for physical and biological sciences. Just as there are parallel discourses of unity and plurality in systems theory, so too are there parallel competing discourses based on differing ontologies in the natural sciences: that of the Newtonian physicists and that of the evolutionary, processual biologists, along with their accompanying modes of ordering and explanation. These threads of interweaving epistemologies and ontologies reinforce and conflict with one another in various complex ways, but for the purpose of exegesis and analysis, we are going to treat them as separate and distinct.

The two characteristics that serve best to discriminate between these intertwining strands are scale and historical time. The question of scale is central to the question of multiplicity. Scale is a form of spatiality, and as Doreen Massey has so clearly argued, there is no space without multiplicity, and no multiplicity without space (2005, 9). Those who find scale-free or scale-invariant ordering, like fractal or network theorists, also find no difference between levels or kinds of systems (e.g., Barabasi and Bonbea 2003). The physical, chemical, neurological, ecological, social, and cultural are all just undifferentiated systems susceptible to the same forms of patterning and analysis. The micro and the macro are similarly indistinct, merely different locations on a continuum.

The emphasis in these scale-invariant forms of ordering in systems is usually on self-organization or autopoiesis (Roth and Schwegler 1990). That is, the system is closed in the sense that its capacity to structure and reproduce itself is independent of external input or feedback. What this implies is a form of causation not normally conceived within the ontology of deterministic entropic systems, which Donald Campbell called "downward causation" (Bickhard and Campbell 2000). In downward causation, the system itself feeds back into the behavior of its parts. It is the concept of downward causation that is the key link to the other discourse of complexity. In this discourse, the emphasis is not so much on self-organization as on emergence, where the question of time or history is paramount.

Emergence is the production of innovative, unanticipated effects that cannot be explained in terms of the properties of the constituent parts of the system. Classic examples are water, whose fluidity and wetness are not properties of its component molecules hydrogen and oxygen, or consciousness, which is not a property of synapses or any of the separate neurophysiological components of the brain (Capra 2003). Like scale, time or history usually discriminates between unified and multiplicitous forms of systems theory.

To revert for the moment to the big metaphysical picture, the prevailing cosmological orthodoxy is the so-called big bang theory, according to which the universe had an origin 13.7 billion years ago, and in an enormously rapid and super-hot explosion, matter, space and time were formed. Although the whole process of the development of the universe as we know it has been given a time line, competing temporal narratives are at play. In the unified physicalist discourse, the laws of physics unfold in a deterministic way, a temporal but essentially ahistorical process in which time is reversible and hence simply expunged. In the processual evolutionary discourse, the development of the universe results from the interaction of irreversible processes subject to selection according to their variable adaptive and connective capacity. The essential difference in the two lies in the irreversibility of time, as Isabelle Stengers has forcefully pointed out, which brings with it multiple temporalities (Stengers 1997; Massey 2005, 129).

There are some profoundly interesting variants of this evolutionary adaptive historical perspective: according to Maturana and Varela (1987), life is a cognitive process; according to Lee Smolin (1997), the cosmos is a multiverse in which universes reproduce through black holes, and our universe, complete with humans and all of nature, has evolved through natural selection. A historical adaptive approach to complexity and systems theory is inherently multiplicitous and experimental; variants are thrown up all the time, and what currently exists and its particular conformation are contingent on what actually happened and what connections were made in the interactions in and between systems.

A fundamental process in complex adaptive systems, especially those invoked by Maturana and Varela, is assemblage. In their view, life is a biological process of connection and interaction whereby life or the system furthers itself. Assemblage or connection is both a form of territorialization or spatial practice that produces multiple distinct wholes and also a cognitive process. Cognition is thus the activity involved in the self-generation and self-perpetuation of networks; it is not a matter of knowing or representing an already independent existing world but rather a continual "bringing forth of a world" through the process of living. "To live is to know" (Maturana and Varela 1987, 28; Capra 2003, 32). For Maturana and Varela, process and structure are the basic phenomena of life. Life is a process of becoming and knowing; it is performative assemblage and movement.

However, to make the processes of assemblage analyzable and visible, it is necessary to hold the narratives of unity and plurality, the arborescent and the rhizomatic, the branching and the random, the smooth and the striated, in tension with one another (Colebrook 2002). This tension is what Deleuze and Guattari address: the problem of living and working with multiplicity in a process of becoming. It requires a new way of writing and thinking that allows the competing narratives and ontologies to interrogate each other so that the hidden spatial and temporal orderings can become apparent, while simultaneously allowing for an attendant reflexive requirement that this approach itself be incorporated in the analysis. It also follows that the standard methods of logical analysis, empirical evidence, and experimental prediction and confirmation will be of limited use; what we also need are the informal insights of experience that Dewey called abduction. As Goldenfeld and Kadanoff put it in *Science* back in 1999: "Up to now physicists have looked for fundamental laws true for all times and all places. But each complex system is different: apparently there are no general laws for complexity. Instead one must reach for 'lessons' that might with insight and understanding be learned in one system and applied to another" (Goldenfeld and Kadanoff 1999, 89).

To allow such an abductive comparison between the functional characteristics of complex adaptive systems and the social production of knowledge, we have brutally summarized in tabular format some of the key points about complex adaptive systems (table 8.1) and about the nature of knowledges and their production (table 8.2).

There are thus strong analogies between the social processes of knowledge production and the components of complex adaptive systems (Marsh

**Table 8.1**
Characterization of complex adaptive systems including functional and ontological dimensions.

---

The overall functional capacity of such systems is for assemblage, connection, and movement with their own creative dynamic, but without any directionality. The following analytic divisions are of necessity somewhat arbitrary, since all components of such systems are interactive and interdependent processes, but together they constitute the essential components.

**Complexity**
1. Multiplicity: there is both variability and diversity within and of systems at every level, allowing for massive redundancy and alternative possible paths.

2. Spatiality: there is an inherent topology where locality matters; there are scale-independent levels and modules whose spatial relations coproduce further niches and spaces.

**Adaptivity**
3. Processuality: the basic ontology is one of processes in continuous states of becoming and interaction, of being selected and reinforced as structures and entities, or abandoned as failures—natural experiments in action.

4. Temporality: the processes are biological and historical, profoundly inflected by the irreversibility of time and the contingency of events, which provide both the dynamic and the diversity.

**Systematicity**
5. Interactive connectivity: There is no prewritten plan, map, logic, algorithm, or laws, no direction or purpose. The structural organization of the system emerges from the interactions of the components and their connections. The multiple parts are agentive in that they do work that creates distinctions or discriminations as elements, states, events, or processes in terms of spatial or temporal relations.

6. Stigmergy: the system is stigmeric; its distributed parts work as a collective through a capacity for tagging that allows information transfer, storage, and processing, and the possibility of negative and positive feedback.

7. Emergence: The system is performative and constructivist; its own connective activities produce systemic spatial and temporal effects and relations that are not in the capacity of the components. These emergent effects reflexively feed back into the components in a process of ecological coproduction. The process whereby the system acquires features that permit it to discriminate, act on, and respond to the environment, its own state, and to other systems is in effect "emergent mapping" and crucially depends on protocols or practices for balancing negative and positive feedback, practices that are themselves not preset but emergent.

---

and Onof 2008). The social production of knowledge is collective work in which we shape ourselves and our environment, and hence it is profoundly spatial, with social, practical, epistemological, and moral dimensions. The cognitive and social order coproduce each other and in so doing construct the kinds of mental and moral spaces we inhabit as knowledge producers. Knowledge and society do not merely interact or determine each other; they

**Table 8.2**
Characterization of knowledges as social processes.

Knowledges are disunited and multiple, collective, and distributed: All knowledges including science and technology are disunified; they are multiplicitous, messy motleys. There are knowledges, not just knowledge singular. There are differing ontologies, methodologies, and epistemologies both within and between cultures. Knowledges do not just inhere in individual minds; they are distributed, created, narrated, practiced, and performed in interaction with other people and things.

Knowledges are tacit, practice based, and embodied: knowledges are not just abstract or just information but skilled practices embodied in the coordinated movements of hands and eyes.

Knowledges are local: they are place based, produced at particular sites by particular people with particular skills, practices, and tools.

Knowledges are mobile, and they travel: universality and unity, supposedly essential characteristics of knowledge, are not in the nature of the knowledges themselves but in the ways that have been developed for moving, circulating, and assembling forms of knowledge.

Knowledges are spatial: making connections, linking people, places, and practices, produces knowledge spaces where trusted agents, significant sites, and traditions are woven together in multiplicitous particular topologies.

Knowledges are socially coproduced and emergent: In making our knowledges of the world, we make ourselves, our societies, and the spaces we inhabit. In moving and acting, we perform our knowledge spaces, we create trails, we know as we go through the cognitive and physical landscape, we are mapping.

Knowledges are performative: Knowledges are the product of human movement, actions, practices, and protocols. It takes active work to transform the world into knowledges, and they are embodied in people, in their practices and relationships, and in their tools, artifacts, and all forms of technology, especially modes of representation, communication, and mensuration. Knowledges are not simply abstract representational entities; they are our conjoint practices and performatively shape the world we inhabit.

Knowledges are narratological and temporal: All knowledges are storied practices. Narratives order events, people, and activities in space and time; in the process of creating meaning and dialogical exchange, they instantiate ontologies.

Knowledges are flexible, dynamic, complex, and negotiated: knowledges in all traditions are produced in flexible, contested, and dynamic processes and practices, but their complexity and negotiated character are erased as they are presented as authorized, accredited, unified public knowledge.

are performatively constitutive of each other. But knowledges, be they social or natural or a combination of the two, do not form anything like coherent wholes. They are complex, messy, and multiple evolving experiments.

Life is a cognitive process, and its apparent coherence, despite the many ways our life experiences are fragmented, disjointed, and disunited, is achieved through our telling stories, taking journeys, marking trails, and making meaning through spatial and temporal ordering (Turnbull 2002a, 2004, 2007; Briggs 1996). Ken Baskin has made explicit the connection between complex adaptive systems, space, and stories in his conception of "storied spaces." For Baskin (2008), "We human animals experience the world in terms of the stories that we believe tell us what 'reality' is, stories that we ourselves co-create as we interact with others in our various social environments—families, organizations, professions, etc.—each of which functions as its own storied spaces. ... These storied spaces function as the human equivalent of complexity's complex adaptive systems."

John Holland (1995, 14) was one of the first to develop a theory of complex adaptive systems. In his analysis, tagging is the key mechanism for aggregation and boundary formation in such systems, enabling differentiation and classification. Tagging, from a coproductive, performative perspective, is a joint effect of the movement of the agents in the system and the interactions between the system and the agents. Tags are the signs of work; they are names, labels, definitions, and indications of interest, value, and concern. Connected together, they form "cognitive trails" that are both the effects and the components of a double mapping process in the coproduction of knowledge and space.[12]

For example, Ochs, Jacoby, and Gonzales (1994) have looked at the ways a group of physicists work collaboratively in reaching an understanding of their complex masses of data, and have found that they take "embodied interpretive journeys" through the representations they share with each other, and in the process create an "intertextual space." Similarly we all make knowledge, our own understandings as we move through space. As Ingold suggests, "We know as we go"; knowledge is cultivated by moving along paths or trails (2000, 228–229). "All knowing is like traveling, like a journey between parts of the matrix" (Turnbull 1991). In moving through space and in making meaning, we are telling stories, making trails of connections, while at the same time those stories and trails, both terrestrial and mental, are the spaces we inhabit.

Claudio Aporta's research reveals a telling example of this interconnection between stories, trails, and constructed spaces in the culture of the Inuit, for whom "moving is a way of living." Although they use no maps and their trails are ephemeral, vanishing with the snow each spring, the Inuit have an enduring network of trails across the Canadian Arctic, a network that endures as stories and shapes their topographic understanding so powerfully that Aporta argues the storied trails are, in effect, places.

The narrative of a journey is not a mere literal description of the trail but involves the story of the trip (and sometimes of different trips along the same route). Such narratives include precise descriptions of the landscape and icescape, along with the memory of personal anecdotes. Place names, winds, and other spatial markers are constantly used to place the traveler within concrete horizons and to explain the direction of travel. The physical description of a trail (including such things as when a particular rock is seen approaching from the trail in a particular direction) is intertwined with stories such as how the traveler almost got lost, the particular hauling of the traveler's father's dogs, the presence and hunting of caribou along the way, or the encountering of another traveler (Aporta 2009, 2003, 2004, 2005).

Complex adaptive systems in nature and in knowledge production are performative in both senses: they capture the ways we know and shape the world, just as our stories and understandings of the world are shaped by our moving through it. Storied spaces and complex adaptive systems are dynamic, dialogical, and diverse ways of knowing and connecting that are in continuous development with emergent effects feeding back into our lives and our environment. Those links can be reinforced by positive feedback or diminished by negative feedback, allowing for the emergent mapping of interwoven stories in a metaspace or third space, a space that is not established by an existing plan, classification system, or assumed ontology.

## Part II: The Shape of the Indigenous Knowledge Commons

### Tall Order

If knowledges are indeed best conceived along the lines suggested in the theoretical discussion summarized in tables 8.1 and 8.2, then what are the creative ICT implications for the world of knowledges: storage and preservation; education and transmission; renewal, growth, and, in the old terminology, discovery? That is the tall order we faced when we attempted to work with the problem of multiple knowledges in practical assignments such as (1) teaching in a tribal college[13] and (2) developing databasing techniques for museum collections containing objects of material culture from around the world.[14]

Both of these projects share the problem at the heart of the tall order: the relationship between ICT and knowledge. Is it possible to sustain genuine diversity of knowledges in any database or digitally mediated mode of knowledge assembly? How is it possible to enable differing ontologies and epistemologies to work together within one system of coordination without subjugating them into a single technologically mediated Western ontology? This requires the kind of radical reconception of the nature of knowledge discussed earlier, as well as the creation of a new kind of "knowledge commons" that does not just archive cultural diversity but uses that diversity dynamically for growth and innovation. In other words, we hope to bring disparate knowledge systems together in a way that produces new knowledge, in an "emergent creative process"—a natural experiment in action.[15]

The advantages of using the World Wide Web to implement the tall order are obvious. As the largest hypertext document in existence, the Web already connects the world's myriad knowledge systems by providing instant linkages, via the Internet, among millions of relevant texts, images, videos, and other multimedia formats. The advent of the Web was an immediate boon to the transmission of indigenous knowledges. In the period of a single generation, most indigenous tribes around the world gained a degree of access to the Internet. In the Americas, even the poorest tribes have some presence there, bringing an end to the Western World Web's monopoly access to the technologies of communication. The Web brought with it the ability to convey the images and sounds of sacred places, rock art, animal calls, songs, dances, elders telling stories in language, indeed, enabling most of the many historical modes of transmitting traditional knowledge. In theory at least, knowledge transmissions on the Internet can be securely directed to specific recipients, or they can be broadcast to the entire world. In short, although the digital divide remains in place for large numbers of indigenous people, access at the tribal level has greatly improved.[16]

### Trails and Storied Spaces in Time

In this final section, we describe the work of the TASSIT Project (Trails and Storied Spaces in Time),[17] a team working to design digital "third spaces"[18] in which the messy, multiplicitous, and incommensurable components of a lifeworld can connect and interact, and in which stories, videos, photos, audio, and scientific texts can be linked by users finding and tagging commonalities and connections, spatially, temporally, and narratologically. Three Web applications and a Web site have resulted from this collaboration so far: Native Trails of Knowing, Story Weaver, and Storied Spaces, each of these applications enabling the construction of a form of knowledge space in which differing traditions are narrated and performed together and in which actors move, make connections, and produce new spaces. The Indigenous Knowledge Commons is a digital bank of resources in support of interdisciplinary teaching and research that bring together both mainstream and indigenous academic traditions. Readers can access descriptions and examples of each of the three Web applications at the Indigenous Knowledge Commons website (http://indigenousknowledge.org).

Story Weaver was developed for specific use in the indigenous liberal studies program at the Institute of American Indian Arts in Santa Fe, New Mexico. Since 2007 students at IAIA have been constructing their own storyweaves as a major course project in several different online courses, including Indigenous Perspectives on Nature, Indigenous Perspectives on Humor, and StoryWeaving: Ways of Knowing and Telling.

Using the Web-based Google Maps software, Native Trails of Knowing is an application designed mainly for instructors preparing study materials in which the knowledge and ideas to be conveyed are pinpointed to specific geographic locations with tags and notes containing text, images, and sound and video files. This will be used in online courses such as American Indian Mapping, Indigenous Perspectives on Knowledge, and How Indians Made America.

The application Storied Spaces (still in the early stages of preparation) constitutes a slightly different approach, suitable for both students and scholars to develop knowledge spaces devoted to intellectual projects that cut across the lines usually drawn between Western science and indigenous knowledge. Links within and between Storied Spaces can be reinforced by positive feedback or diminished by negative feedback, allowing for the emergent mapping of storyweaves in a metaspace, or third space, a space that is not established by a preexistent plan, classification system, or assumed ontology. We hope that the Storied Spaces facility will be useful both for teaching at the tertiary level and as an adjunct to collections data-basing systems at museums.

In the following paragraphs, we describe (in words only, of course) some of the content of specific knowledge spaces developed by the TASSIT Project (which could be presented either in Story Weaver or in Storied Spaces). These spaces attempt to recontextualize accounts that may have first appeared as scholarly accounts, elders' stories, spiritual journeys, or social interactions.

### Listening to Stones and Glaciers

Almost universally found as a central tenet of indigenous knowledges around the world is a perspective that is, by any standard, difficult or impossible to reconcile with Western science: the idea that animals, plants, rocks, and humans "are all related," that all living things, and even stones or clouds, may be sentient beings that have consciousness and can "speak" to those who listen. Linda Tuhiwai Smith describes the Maori knowledge (*whakapapa*) as a "way of thinking, a way of learning, a way of storing knowledge, and a way of debating. ... Whakapapa also relates to all other things that exist in the world. We are linked through our whakapapa to insects, fishes, trees, stones, and other life forms." Gregory Cajete (Tewa Pueblo) considers indigenous knowledge to be a "natural democracy in which humans are related and interdependent with plants, animals, stones, water, clouds and everything else" (Smith 2000, 234–235; Cajete 2004, 46; Roberts et al. 2004). Similarly, George Tinker (Osage) develops this idea further in a philosophical essay:

> Did you know that rocks talk? Well, they do. Yes, I am aware that this is an audacious claim—even for an American Indian—made in the context of late modernity (or even postmodernity, if you insist) and in the context of a world indelibly marked by the accomplishments of modern science. But the argument proposed for this essay is that rocks talk and have what we must call consciousness. And we must extend our discussion of rocks to trees—as Walking Buffalo asserts—and to the rest of the created world around us. (Tinker 2004)

There are, of course, a number of levels of interpretation of such indigenous beliefs, from literal to spiritual to metaphoric to ecological. Clara Sue Kidwell (Choctaw/Chippewa) observes:

> Attuned to the rhythms of that environment, keenly aware of its vicissitudes, and dependent upon their relationships with the animals and plants, Native American societies understood the basic spirituality of the world. ... The realm of the spiritual is that realm where something in the environment—a rock, a plant, a storm—exhibits behavior that is unpredictable. Suddenly, a rock is alive, a plant has special powers to cure a particular physical condition, or an old woman walking down a path suddenly disappears, to be replaced by a bear. (Kidwell 2003, 1043)

Julie Cruikshank's (2005) "Do Glaciers Listen?" constitutes a remarkable example of how indigenous knowledges can be treated credibly without subverting them and without losing the capacity for criticism. This is a brilliant exposition of the radically different ways in which glaciers are understood by scientists and the Tlingit and Athapaskan peoples. The indigenous people of the Alaskan panhandle have an ontology that radically disrupts the nature–culture divide that the West and its sciences present as self-evident. For them, glaciers are animate; they taste, smell, take action, and make moral judgments. Cruikshank makes the useful distinction between "listening to a story" and "listening for a story" (which she attributes to Eudora Welty) and is able to hold in tension the different narratives and performances of makers of knowledge of glaciers, both indigenous and scientific, without ceding superiority to either. Her book is, in itself, a wonderful "storyweave" of maps, impeccably presented stories by elders, and glaciology, along with historical and sociopolitical dimensions.

A final element in this storied space comes from the work of Kathryn Geurts (2002), who shows that differing knowledge traditions can have profoundly different sensory bases (see also Classen 1993). The Anlo-Ewe in Ghana articulate a sixth sense in understanding that is profoundly involved with the society's epistemology. It is the sense of proprioception, of balance, a sense that goes largely unrecognized and unacknowledged in the West. It is a way of knowing that again disrupts the idea that the only valid knowledge is empirical, warranted by the five senses.

All the elements just discussed are taken together as a colloquy of voices, enhanced in Story Weaver or Storied Spaces, with videos of surging glaciers, grandfather stones, songs, animal calls, vivid natural images, and perhaps even kinesthetic manifestations of knowing. Such a storied space supplies some of the tacit knowledge hinted at but impossible to convey in a printed text. Insights like these move closer to where an answer to all these questions about knowledges and ICT may be located: interactive digital spaces.

**Termites and Cathedrals**
The requirements of an indigenous knowledge commons for diversity without centralized coordination are well illustrated by a story about termites and cathedrals, about knowledge and space and how they are coproduced through movement and interaction. In northern Australia termite mounds are called "cathedral mounds," presumably because of their superficial structural resemblance. But the more profound association between termites and cathedrals is reflected in recent work on the ways in which termites use stigmergic techniques to construct their mounds, follow trails to create the shortest routes to food, and manage their community at optimal effectiveness (Bonabeau, Dorigo, and Theraulaz 1999; Theraulaz and Bonabeau 1999; Theraulaz et al. 2003).

The concept of stigmergy was originally proposed by Grassé in the 1950s and is now finding widespread application in distributed systems from AI to digital ecosystems. Stigmergy, a neologism coming from stigma sign and ergon work, captures a basic principle that seems to explain "the coordination paradox" underlying distributed systems (Elliott 2006; 2007, chap. 3; Theraulaz and Bonabeau 1999). How is the ad hoc collective, collaborative work of thousands of individual agents such as termites spatially and temporally coordinated to build complex structures like termite mounds? The problem being that there is no evidence of a plan, or a group mind, or of individual termites having any sense of the larger enterprise. So how do the termites know what to do? The basic stigmergic principle, as formulated by the Belgian cyberneticist Francis Heylighen, goes as follows: "Work performed by an agent leaves a trace in the environment or medium, perceiving the trace stimulates another agent to perform further work, thus extending or elaborating previous work."[19] In the case of termites, the effect can be either active and direct (so-called sematectonic stigmergy), where changes to the environment induce further changes, or passive and indirect (so-called marker stigmergy), where the marker does not itself have any function (Karsai 1999).

In developing the shortest path to food, termites randomly search, leaving trails of pheromones that slowly evaporate. The shortest path is reinforced through the indirect stigmergic feedback mechanisms of pheromone renewal by other that termites follow the path, while other paths fade out as they are not followed repetitively. The reinforced path acts as an external memory, a cognitive trail, a knowledge store, a positive feedback that importantly does not require a complex mechanism or even "information transfer." Similarly, in building the mound, the termites secrete mud balls marked with pheromones and deposit them where there are other mud balls. These piles of secreted "tagged" balls rise into towers, and as the lower pheromone tags fade away, the towers start leaning toward other towers, thus forming arches. Eventually about five cubic meters of earth is turned into a vertical cooling system, driven by the increasing wind pressure at higher levels above the ground and oriented by the Earth's magnetic field, allowing minimization of solar exposure (Turner 2002).[20]

While it is easy to see that termites do not have a plan, a design, or an architect to direct operations, it is less obvious that the early gothic cathedrals were built by the same ad hoc collective and collaborative work of many men. The work of the masons, carpenters, and thousands of others involved in building the early cathedrals like Chartres was not coordinated through plans, as we now expect, but through talk, templates, and tradition (Turnbull 2003). Plans in the sense of detailed working drawings were developed as a technology in the overall process of the gothic cathedral-building crusade and were initially introduced as a form of archiving or recording the building process. Only later were plans adopted as the essential technique for innovation and construction, eventually becoming a hallmark of modernity. Now, it is plausible to see the clearly cultural activity of cathedral building and the clearly biological activity of termite mound building as being socially distributed cognition in action and as having stigmergic processes in common.

Indeed, in Africa we see examples of complex termite-human interactive processes. "Local communities have comprehensive indigenous knowledge of termite ecology and taxonomy ... and also have elaborate knowledge of the nutritional and medicinal value of termites and mushrooms associated with termite nests. ... Subsistence farmers use termites as indicators of soil fertility and use termite mound soil ... for crop production" (Sileshi et al. 2009). In North Australia didgeridoos made of eucalyptus trunks hollowed out by termites are among the oldest wind instruments in the world. Typically decorated with complex designs, they are themselves a story, and through them Aborigines sing and tell stories, perform ceremonies, and connect to the Dreaming. Didgeridoos are a key device for transferring knowledge across the generations.

Although the history of adopting biological metaphors has been disastrous, these two storyweaves reveal deep similarities between the biological and cultural conceptions of knowledge creation and innovation. Biology is emerging as "the" science of the era, dominated, as it is, by the profoundly deterministic sciences of genetics and molecular biology. However, this rise to dominance has been accompanied by a more thoroughly historical, evolutionary, and performative conception of biological processes as natural experiments, and this conception lies at the heart of the new thinking about complex adaptive systems as storied spaces or as distributed interactions in evolving networked communities (Woese 2004; Reid 2007).

## Notes

1. Many universities now think of their libraries as a knowledge commons. In South Africa, for instance, CSIR has established a knowledge commons.

2. India's CSIR challenged the patent for turmeric issued to the University of Mississippi Medical Center in December 1993. As a result of the challenge, the U.S. Patents and Trademarks Office ruled on August 14, 1997, that the patent was invalid because it was not a novel invention. See http://www.rediff.com/news/aug/27haldi1.htm.

3. On the status of indigenous knowledges in India, see the Traditional Knowledge Digital Library (TKDL), http://www.tkdl.res.in/tkdl/langdefault/common/home.asp?GSL=Eng; http://www.ciks.org. For Africa see the World Bank, http://www.worldbank.org/afr/ik/achieve.htm.

4. Indigenous Knowledge Systems (IKS), http://www.nrf.ac.za/projects.php?pid=45.

5. BBC News, May 30, 2008, http://news.bbc.co.uk/2/hi/americas/7426794.stm.

6. Thaindian News, July 18, 2008, http://www.thaindian.com/newsportal/sports/the-amazon-tribe-that-speaks-language-without-numbers_10073103.html.

7. See also Bowker's (2000) "dynamic uncompromise." Zia Sardar (2006, 254–256) recently called for a recognition of the transmodern.

8. See http://www.theguardian.com/environment/2009/mar/24/george-monbiot-climate-change-biochar.

9. Yoffee and MacAnany (2009) have gone some way toward setting the record straight but do not provide for a fresh understanding of IK.

10. For a useful short overview focusing on emergence, see Corning 2002.

11. See many of the authors in *Cycles of Contingency: Developmental Systems Theory and Evolution* (Oyama, Griffiths, and Gray 2001b).

12. The term "cognitive trail" is Adrian Cussins's, who argues that a "traveling account of understanding and representation should not opt for an epistemological grounding in thought or experience," since much of our "intelligence in communicating and acting consists in our ability to *move between* alternative conceptualizations of a problem domain" (Cussins 1992, cited in Turnbull 2002b, 135). Trails, as Cussins points out, are also artifacts. "Perhaps trails are the *first* artifacts" (Cussins 1997, 14).

13. The Native Eyes Project at the Institute of American Indian Arts (IAIA), under the leadership of Wade Chambers (Cherokee descent, Ph.D., history of science), was founded as an innovative program of online courses designed to incorporate social, cultural, and intellectual contributions of indigenous peoples into mainstream teaching in the humanities, sciences, and social sciences. Consultants to the project have included Dave Warren (Santa Clara Pueblo, Ph.D., history), Nancy Mithlo (Chiricahua Apache, Ph.D., anthropology), David Turnbull (Ph.D., history and philosophy of science), Laurie Whitt (Mississippi Choctaw descent, Ph.D., philosophy), Greg Cajete (Santa Clara Pueblo, Ph.D., education), Carlos Andrade (Kaua'i, Ph.D., geography), LaDonna Harris (Comanche), and Jim Enote (Zuni, director of A:shiwi A:wan Museum). Consultant museums include the Museum of the American Indian, the Peabody Essex Museum, the Canadian Museum of Civilizations, and the Museum of Contemporary Native Arts.

14. Along with a group including Geof Bowker (cyberscholarship, Pittsburgh), Leigh Star (library and information science, Pittsburgh), Michael Bravo (Scott Polar Research Institute), and Ramesh Srinivasan (information studies, UCLA), we have worked with Robin Boast (information science and culture, University of Amsterdam) on rethinking the database of the museum's collection of artifacts. That project, $E^2D^2$ (Emergent Databasing, Emergent Diversity), aims to avoid, as the name suggests, giving the database a fixed and preestablished classification system or ontology. At the same time, it aims to incorporate the understandings and ontologies of the cultures that produced the materials in their collections, and to allow the interrogations and interactions of the database by its users and contributors to produce emergent practices and knowledge.

15. Emergence is discussed later, but the notion of emergent practices as the way to work with incommensurability was suggested by Donna Haraway (2003, 7), drawing

# Assembling Diverse Knowledges

on Helen Verran's notion of emergent ontologies. "How can people rooted in different knowledge practices 'get on together,' ... How can general knowledge be nurtured in postcolonial worlds committed to taking difference seriously? Answers to these questions can only be put together in emergent practices; i.e., in vulnerable on the ground work that cobbles together non-harmonious agencies and ways of living that are accountable to both their disparate inherited histories and to their barely possible absolutely necessary joint futures." See also Pickering 1995; Pickering and Guzick 2008; and Pickering's notion of emergence in the "dance of agency."

16. In this limited space, we are not able to treat many weighty questions highly relevant to this discussion: issues of intellectual property, hierarchies of authority and identity, the digital divide, security in relation to secret sacred knowledge, and more.

17. In 2006 the authors approached the Web design firm Inventive Labs to explore the possibility of developing a series of Web applications for this purpose.

18. If Western science and indigenous knowledges have their own spaces within which they speak with authority and privilege, we aim to construct third spaces in which differing traditions are narrated and performed together and in which actors can move, make connections, and produce new spaces and trails (Turnbull 2002b).

19. Francis Heylighen, *Stigmergy: A Fundamental Paradigm for Digital Ecosystems?* http://www.digital-ecosystems.org/de-2007/speakers1.html.

20. Turner makes the important point that stigmergy is largely positive feedback and if left uninterrupted would fill the mound in. He finds the source of social disruption in the fungi that the termites cultivate, which occasionally throw up mushrooms that burst through and disrupt the mound, leading him to pose the question, who cultivates whom? See Turner, "Symbiotic Fungi and Social Homeostasis," http://www.esf.edu/efb/turner/old%20site/termite/fungi.htm.

## References

Aporta, Claudio. 2003. New ways of mapping: Using GPS mapping software to plot place names and trails in Igloolik (Nunavut). *Arctic* 56 (4): 321–327.

Aporta, Claudio. 2004. Routes, trails, and tracks: Trail breaking among the Inuit of Igloolik. *Études Inuit Studies* 28 (2): 9–38.

Aporta, Claudio. 2005. From map to horizons; from trail to journey: The challenges of documenting Inuit geographic knowledge. *Études Inuit Studies* 29 (1–2): 221–231.

Aporta, Claudio. 2009. The trail as home: Inuit and their Pan-Arctic network of routes. *Human Ecology* 37:131–146.

Bak, Per. 1996. *How Nature Works: The Science of Self-Organized Criticality.* New York: Springer.

Balée, William, and Clark Erickson. 2006. Time, complexity, historical ecology. In *Time and Complexity in Historical Ecology: Studies in the Neotropical Lowlands*, ed. William Balée and Clark Erickson, 1–20. New York: Columbia University Press.

Barabasi, Albert-Laszlo. 2002. *Linked: The New Science of Networks*. Cambridge, MA: Perseus.

Barabasi, Albert-Laszlo, and E. Bonbea. 2003. Scale-free networks. *Scientific American* 289 (5): 50–59.

Baskin, Ken. 2008. Storied spaces: The human equivalent of complex adaptive systems. *E:CO* 10 (2): 1–12.

Bates, Peter, Moe Chiba, Sabine Kube, and Douglas Nakashima. 2009. *Learning and Knowing in Indigenous Societies Today*. Paris: UNESCO.

Bertalanffy, Ludwig von. 1968. *General Systems Theory: Foundations, Development, Applications*. New York: George Braziller.

Bickhard, Mark, and Donald Campbell. 2000. Emergence. In *Downward Causation: Minds, Bodies, and Matter*, ed. Peter Andersen, Claus Emmeche, Niels Finnemann, and Peder Christiansen, 322–348. Aarhus: Aarhus University Press.

Blaser, Mario. 2009. The threat of the Yrmo: The political ontology of a sustainable hunting program. *American Anthropologist* 111 (1): 10–20.

Boccara, Guillaume. 2006. The brighter side of the indigenous renaissance (part 1). *Nuevo Mondo Mondos Nuevos*, June 16. http://nuevomundo.revues.org/index2405.html.

Bonabeau, Eric, Marco Dorigo, and Gary Theraulaz. 1999. *Swarm Intelligence: From Natural to Artificial Systems*. New York: Oxford University Press.

Bowker, Geoffrey. 2000a. Mapping biodiversity. *International Journal of GIS* 14 (8): 739–754.

Bowker, Geoffrey. 2000b. Biodiversity dataversity. *Social Studies of Science* 30 (5): 643–684.

Briggs, Charles, ed. 1996. *Disorderly Discourse: Narrative, Conflict, and Inequality*. New York: Oxford University Press.

Buchanan, Mark. 2002. *Nexus: Small Worlds and the Groundbreaking Science of Networks*. New York: W. W. Norton.

Cajete, Gregory. 2004. Philosophy of native science. In *American Indian Thought*, ed. Anne Waters. Malden: Blackwell.

Capra, Fritof. 2003. *The Hidden Connections: A Science for Sustainable Living*. London: Flamingo.

Chambers, David Wade. 2008. American Indian knowledge transmission. In *Encyclopaedia of Science, Technology, and Medicine in Non-Western Cultures*, ed. Helaine Selin. Dordrecht: Kluwer Academic.

Chambers, David Wade, and Richard Gillespie. 2000. Locality in the history of science: Colonial science, techno-science, and indigenous knowledge. *Osiris* 15:221–240.

Classen, Constance. 1993. *Worlds of Sense: Exploring the Senses in History and across Cultures*. London: Routledge.

Colebrook, Clare. 2002. *Understanding Deleuze*. Sydney: Allen & Unwin.

Corning, Peter. 2002. The re-emergence of "emergence": A venerable conception in search of a theory. *Complexity* 7 (6): 18–30.

Cruikshank, Julie. 2005. *Do Glaciers Listen? Local Knowledge, Colonial Encounters, and Social Imagination*. Vancouver: UBC Press.

Cussins, Adrian. 1992. Content, embodiment, and objectivity: The theory of cognitive trails. *Mind* 101:651–688.

Cussins, Adrian. 1997. Norms, networks, and trails: Relations between different topologies of activity, kinds of normativity, and the new weird metaphysics of actor network theory, and some cautions about the contents of the ethnographer's toolkit. http://www.academia.edu/3449801/Norms_networks_and_trails.

de la Cadena, Marisol. 2006. The production of other knowledge and its tensions: From Andeanist anthropology to *Interculturalidad*? In *World Anthropologies: Disciplinary Transformations within Systems of Power*, ed. Gustavo Ribeiro and Arturo Escobar, 201–224. Oxford: Berg.

Denevan, William. 2001. *Cultivated Landscapes of Native Amazonia and the Andes*. Oxford: Oxford University Press.

Diamond, Jared. 2005. *Collapse: How Societies Choose to Fail or Survive*. London: Allen Lane.

Dussel, Enrique. 1993. Eurocentrism and modernity. *Boundary 2* 20 (3): 65–76.

Dussel, Enrique. 1996. *The Underside of Modernity: Apel, Ricoeur, Rorty, Taylor, and the Philosophy of Liberation*. New York: Humanities Press.

Dussel, Enrique. 2009. Transmodernity and interculturality: An interpretation from the perspective of philosophy of liberation. http://www.enriquedussel.com/txt/Transmodernity%20and%20Interculturality.pdf.

*Economist*. 2004. Indigenous people in South America: A political awakening . *Economist*, February 19. http://www. economist.com/node/2446861.

Elliott, Mark. 2006. Stigmergic collaboration: The evolution of group work. *M/C Journal* 9 (2).

Elliott, Mark. 2007. Stigmergic collaboration: A theoretical framework for mass collaboration. Ph.D. thesis, Centre for Ideas, Victorian College of the Arts, University of Melbourne.

Erickson, Clark. 2003. Historical ecology and future explanations. In *Amazonian Dark Earths: Origins, Properties, Management*, ed. J. Lehmann, D. Kern, B. Glaser, and W. Woods, 455–500. Dordrecht: Kluwer.

Evans, Nicholas. 2009. *Dying Words: Endangered Languages and What They Have to Tell Us*. Chichester: Wiley.

Galison, Peter, and David Stump, eds. 1996. *The Disunity of Science: Boundaries, Contexts, and Power*. Stanford: Stanford University Press.

German, Laura. 2003. Ethnoscientific understandings of Amazonian dark earths. In *Amazonian Dark Earths: Origins, Properties, Management*, ed. Johannes Lehmann, Dirse C. Kern, Brund Glaser, and William I. Woods, 179–204. Dordrecht: Kluwer.

Geurts, Kathryn. 2002. *Culture and the Senses: Bodily Ways of Knowing in an African Community*. Berkeley: University of California Press.

Gleick, James. 1987. *Chaos: Making a New Science*. London: Viking Penguin.

Goldenfeld, Nigel, and Leo P. Kadanoff. 1999. Simple lessons from complexity. *Science* 284 (April 2): 87–89.

Green, Lesley. 2008. Africa's indigenous knowledge systems policy. *Anthropology Southern Africa* 31 (1–2): 48–59.

Green, Lesley. 2009. Knowledge contests, South Africa. *Anthropology Southern Africa* 32 (1–2):2–7.

Gribbin, John. 2004. *Deep Simplicity: Chaos, Complexity, and the Emergence of Life*. London: Allen Lane.

Haraway, Donna. 2003. *The Companion Species Manifesto: Dogs, People, and Significant Otherness*. Chicago: Prickly Paradigm Press.

Harrison, K. David. 2007. *When Languages Die: The Extinction of the World's Languages and the Erosion of Human Knowledge*. New York: Oxford University Press.

Hesse, Charlotte, and Elinor Ostrom, eds. 2007. *Understanding Knowledge as a Commons: From Theory to Practice*. Cambridge, MA: MIT Press.

Holland, John. 1995. *Hidden Order: How Adaptation Builds Complexity*. Reading, MA: Helix Books.

Horgan, John. 1995. From complexity to perplexity. *Scientific American* 272 (June): 74–79.

Ingold, Tim. 2000. *The Perception of the Environment: Essays in Livelihood, Dwelling, and Skill*. London: Routledge.

Johnson, Steven. 2001. *Emergence: The Connected Lives of Ants, Brains, Cities, and Software*. London: Allen Lane.

Karsai, István. 1999. Decentralized control of construction behavior in paper wasps: An overview of the stigmergy approach. *Artificial Life* 5 (2): 117–136.

Kidwell, Clara Sue. 2003. Sacred clowns. In *Encyclopedia of World Environmental History*, ed. Shepard Krech, J. R. McNeill, and Carolyn Merchant. New York: Routledge.

Latour, Bruno. 1993. *We Have Never Been Modern*. Cambridge, MA: Harvard University Press.

Latour, Bruno. 2004. *Politics of Nature: How to Bring the Sciences into Democracy*. Trans. C. Porter. Cambridge, MA: Harvard University Press.

Lewin, Roger. 1992. *Complexity: Life at the Edge of Chaos*. New York: Macmillan.

Maclaurin, J. 2012. Universal Darwinism: Its scope and limits. In *Rationis Defensor: Essays in Honour of Colin Cheyne*, ed. J. Maclaurin. New York: Springer.

Maffi, Luisa. 2005. Linguistic, cultural, and biological diversity. *Annual Review of Anthropology* 34:599–617.

Mann, Charles. 2000. Earthmovers of the Amazon. *Science* 287 (5454): 786–789.

Mann, Charles. 2005. *1491: New Revelations of the Americas before Columbus*. New York: Alfred Knopf.

Marsh, Leslie, and Christian Onof. 2008. Stigmergic epistemology, stigmergic cognition. *Cognitive Systems Research* 9 (1–2): 136–149.

Massey, Doreen. 2005. *For Space*. London: Sage.

Maturana, Humberto, and Francisco Varela. 1987. *The Tree of Knowledge: The Biological Roots of Human Understanding*. Boston: New Science Library.

Nanda, Meera. 2003. *Prophets Facing Backwards: Postmodern Critiques of Science and Hindu Nationalism in India*. New Brunswick, NJ: Rutgers University Press.

Ochs, Elinor, Sally Jacoby, and Patrick Gonzales. 1994. Interpretive journeys: How physicists talk and travel through graphic space. *Configurations* 2 (1): 151–171.

Odling-Smee, John, Kevin Laland, and Marcus Feldman. 2003. *Niche Construction: The Neglected Process in Evolution*. Princeton: Princeton University Press.

Oyama, Susan, Paul Griffiths, and Russell Gray. 2001a. Introduction: What is developmental systems theory? In *Cycles of Contingency: Developmental Systems Theory and Evolution*, ed. Susan Oyama, Paul Griffiths, and Russell Gray, 1–11. Cambridge, MA: MIT Press.

Oyama, Susan, Paul Griffiths, and Russell Gray, eds. 2001b. *Cycles of Contingency: Developmental Systems Theory and Evolution*. Cambridge, MA: MIT Press.

Pickering, Andrew. 1995. *The Mangle of Practice: Time, Agency, and Science*. Chicago: University of Chicago Press.

Pickering, Andrew, and Keith Guzick, eds. 2008. *The Mangle in Practice: Science, Society, and Becoming*. Durham: Duke University Press.

Polanyi, Michael. 1958. *Personal Knowledge: Towards a Post-critical Philosophy*. London: Routledge and Kegan Paul.

Povinelli, Elizabeth. 2001. Radical worlds: The anthropology of incommensurability and inconceivability. *Annual Review of Anthropology* 30:319–334.

Reid, Robert. 2007. *Biological Emergences: Evolution by Natural Experiment*. Cambridge, MA: MIT Press.

Ribeiro, Gustavo, and Arturo Escobar. 2006. World anthropologies: Disciplinary transformations within systems of power. In *World Anthropologies: Disciplinary Transformations within Systems of Power*, ed. Gustavo Ribeiro and Arturo Escobar, 1–28. Oxford: Berg.

Roberts, Mere, Brad Haami, Richard Benton, Terre Satterfield, Melissa Finucane, Mark Henare, and Manuka Henare. 2004. Whakapapa as a Maori mental construct: Some implications for the debate over genetic modification of organisms. *Contemporary Pacific* 16 (1): 1–28.

Roth, Gerhard, and Helmut Schwegler. 1990. Self-organisation, emergent properties, and the unity of the world. In *Selforganization: Portrait of a Scientific Revolution*, ed. Wolfgang Krohn, Gunter Küppers, and Helga Nowotny, 36–50. Dordrecht: Kluwer.

Sandor, J. A., N. Barrera-Bassols, A. Winkler Prins, and J. Zinck. 2006. The heritage of soil knowledge among the world's cultures. In *Footprints in the Soil: People and Ideas in Soil History*, ed. Benno P. Warkentin, 43–84. Oxford: Elsevier.

Sardar, Zia. 2006. *How Do You Know? Reading Zia Sardar on Islam, Science, and Cultural Relations*. London: Pluto Press.

Sileshi, Gudeta W., Philip Nyeko, Phillip O. Y. Nkunika, Benjamin M. Sekematte, Festus K. Akinnifesi, and Oluyede C. Ajayi. 2009. Integrating ethno-ecological and scientific knowledge of termites for sustainable termite management and human welfare in Africa. *Ecology and Society* 14 (1). http://www.ecologyandsociety.org/vol14/iss1/art48/.

Smith, Linda. 2000. Kaupapa Maori research. In *Reclaiming Indigenous Voice and Vision*, ed. Marie Battiste. Vancouver: UBC Press.

Smolin, Lee. 1997. *The Life of the Cosmos*. London: Phoenix.

Stengers, Isabelle. 1997. *Power and Invention: Situating Science*. Minneapolis: University of Minnesota Press.

Theraulaz, Guy, and Eric Bonabeau. 1999. A brief history of stigmergy. *Artificial Life* 5 (2): 97–116.

Theraulaz, Guy, Jacques Gautrais, Scott Camazine, and Jean-Louis Deneubourg. 2003. The formation of spatial patterns in social insects: From simple behaviours to complex structures. *Philosophical Transactions of the Royal Society of London, Series A: Mathematical and Physical Sciences* 361:1263–1282.

Tinker, George "Tink." 2004. The stones shall cry out: Consciousness, rocks, and Indians. *Wicazo Sa Review* 19 (2): 105–125.

Turnbull, David. 1991. *Mapping the World in the Mind: An Investigation of the Unwritten Knowledge of the Micronesian Navigators*. Geelong: Deakin University Press.

Turnbull, David. 2000. Rationality and the disunity of the sciences. In *Mathematics across Cultures: The History of Non-Western Mathematics*, ed. Helaine Selin, 37–54. Dordrecht: Kluwer Academic.

Turnbull, David. 2002a. Travelling knowledge: Narratives, assemblage, and encounters. In *Instruments, Travel, and Science: Itineraries of Precision from the Seventeenth to the Twentieth Century*, ed. Marie-Noelle Bourget, Christian Licoppe, and H. Otto Sibum, 273–294. London: Routledge.

Turnbull, David. 2002b. Performance and narrative, bodies and movement, in the construction of places and objects, spaces and knowledges: The case of the Maltese megaliths. *Theory, Culture, and Society* 19 (5–6): 125–143.

Turnbull, David. 2003. *Masons, Tricksters, and Cartographers: Comparative Studies in the Sociology of Scientific and Indigenous Knowledge*. 2nd ed. London: Routledge.

Turnbull, David. 2004. Narrative traditions of space, time, and trust in court: *Terra Nullius*, "wandering," the Yorta Yorta native title claim, and the Hindmarsh Island Bridge controversy. In *Expertise in Regulation and Law*, ed. Gary Edmond, 166–183. Aldershot: Ashgate.

Turnbull, David. 2007. Maps, narratives, and trails: Performativity, hodology, distributed knowledge in complex adaptive systems—an approach to emergent mapping. *Geographical Research* 45 (2): 140–149.

Turnbull, David. 2009. Introduction. *Futures: Special Issue on the Futures of Indigenous Knowledges* 41 (1): 1–5.

Turner, J. Scott. 2002. A superorganism's fuzzy boundaries: The breathing termite mounds of southern Africa raise the question, where does "animate" end and "inanimate" begin? *Natural History*, July–August.

United Nations Permanent Forum on Indigenous Issues. 2008. Indigenous languages. http://www.un.org/esa/socdev/unpfii/documents/Factsheet_languages_FINAL.pdf.

Vince, Gaia. 2009. Interview: James Lovelock. *New Scientist* 201 (January 24): 30–31.

Viveiros de Castro, Eduardo. 1998. Cosmological deixis and Amerindian perspectivism. *Journal of the Royal Anthropological Institute*, n.s., 4:469–488.

Viveiros De Castro, Eduardo. 2004. The transformation of objects into subjects in Amerindian ontologies. *Common Knowledge* 10 (3): 463–484.

Waldrop, M. Mitchell. 1992. *Complexity: The Emerging Science at the Edge of Order and Chaos*. London: Penguin.

Watts, Duncan J. 2003. *Six Degrees: The Science of a Connected Age*. New York: W. W. Norton.

Woese, Carl. 2004. A new biology for a new century. *Microbiology and Molecular Biology Reviews* 68 (2): 173–186.

Yoffee, Norman, and Patricia MacAnany, eds. 2009. *Questioning Collapse: Human Resilience, Ecological Vulnerability, and the Aftermath of Empire*. Cambridge: Cambridge University Press.

# 9 Structuring the Social: Inside Software Design

Alan F. Blackwell

The goal of this chapter is to present a view from the inside, as it were. I am concerned with the inside of ICT—because where there is a technology, there must also be technologists. How, then, are the themes of this book perceived by information and communications technologists? It is important to note that the phrase "information and communications technologist" is not one that I would choose myself. Indeed, from within our profession, the phrase "ICT" is often considered pejorative and even hostile, a construction of commentators and policy makers who wish to caricature our work. At the time of writing, the United Kingdom marks the end of a long campaign to remove "ICT" from the British school syllabus, where it is widely regarded as having damaged the education of our children and the economic prospects of the British technology industry (Crookes 2012). In its place, "computer science" or "computing" is being restored as an academic subject—in this case, names chosen and recognized by the field's practitioners.

This volume offers a fascinating opportunity to see indigenous communities from the inside. As with all anthropological research, colonial knowledge systems (whether scientific or commercial) are challenged by participating in native cultures, observing their practices, and listening to their descriptions. The assumptions of ICT are certainly challenged by closer engagement with the social contexts in which technology is used by indigenous people. It may also be the case that social science is challenged by engagement with the ways that social science is used within the native cultures of ICT. The purpose of this chapter is to describe and reflect on that encounter, as framed by the wider concerns of the present volume.

The plan is first to describe the distinctive nature of computing as a profession, shaped by its technical concerns within a historical context of professionalism. I then address the engagement that the profession has with social contexts, and the ways in which social science has become a key element of this engagement. A specific issue of concern is the relationship between the design of commercial products, and design research in academic and scientific contexts. This is problematic in large part because the distinctions between the two are not always clear to the publics that we engage with, or to the social scientists with whom we collaborate. Finally, I discuss the ways in which these dynamics played out in a technology research program spanning multiple individual projects, all of which had the objective of engaging with people in developing nations. As director of the funding program, I was responsible both for framing and reflecting on the role played by social scientists in interdisciplinary design research.

**Structure and Materiality**

Above all else, software is structure. Even "social software" revolves around engineered structure, whether the fantasies of a Harvard undergraduate for a reference book listing the pretty faces in his classes (Facebook), or popular opinion captured via 140 character nodes in a data-mining graph (Twitter). In essence, the task of the software professional is to create structured accounts of states of affairs, and then to construct structured interventions mediated by technology. As with any technology, structure in software is constructed both explicitly, via engineering principles and procedures, and also tacitly, via craft practices that are embodied in tools and practitioners (Woolford et al. 2010; Blackwell 2013). Technologists often refer to the conjunction of engineering and craft simply as "design," although the conflation of engineering and craft traditions in this term draws attention to the tensions of economic and caste relations between a (middle-class) "profession" and (working-class) "trade."

During the agricultural and industrial revolutions, these tensions and dynamics were usually embedded in, obscured by, or revealed through the materiality of the technological artifacts and their contexts of production and consumption. However, the immateriality of information technologies—of their software parts and relations—means that physical artifacts are no longer the preeminent basis of structure in design. Although well understood by software technologists themselves, this fact is intentionally obscured by the manufacturers of ICT products and infrastructure, and by the consumerist media commentary that drives the frenzy accompanying each new gadget. As a result, design is falsely characterized as a conjunction of consumerist brand loyalty and romantic hero-creators. The real practice of ICT design may be equally consumerist, exploitative, Westernized, and so on, but an understanding of this practice must be informed by an understanding of software itself as a technology of structure.

I therefore focus here on the specificity of software design, as an engineering/craft practice, that is informed both by reflection on the social context of production and by observations of the social context of use. Engineering practices continually appropriate research methods and analytic perspectives from the sciences, and here the methods and theories of the social sciences become a resource for engineering design. Reflective craft practices also seek critical vocabularies for their work, as well as novel sources of insight into the nature of the problems they address. For craft designers, ethnographic practices and social critique are therefore reshaped as new additions to the craft toolbox. However, it is essential that "engineering" and "craft" are here understood in their new forms, emphasizing the immateriality of the software artifact, and its consequent preeminent concern with structure. Design researchers (whether from engineering or craft tendencies) have professed openness to engagement with social science in the academic and professional venues associated with interface design, interaction design, and experience design. In the context of this book, my main concern is to measure that engagement with social science, in software design contexts, against the broader questions raised by the theme of cross-cultural knowledge encounters and the politics of design.

**The Elision of "Design"**

It is necessary to distinguish software design from computer science. Although both deal with immaterial structure, software design, like all design, is a set of situated professional practices. To the extent that these practices are formalized, they are described as "methodologies." Computer science is an academic discipline, arranged in the conventional manner, whose relationship to software design reflects the conventional relations of professional certification. The descriptions of ICT development in this book are descriptions neither of computer science nor of software design, but there are many cases in which the differing practices and commitments of "science" and "design" are elided, with the two treated more uniformly as general and homogeneous representations of colonial, corporate, nonindigenous knowledge systems.

It seems possible that in an undifferentiated critique of conventional science, the actual professional practices of ICT have been replaced by those of science more generally, to an extent that the encounters with computing professionals have unfortunately failed to capture either their professional concerns or epistemological commitments. I turn now to addressing these questions through encounters of software designers with social scientists and with development contexts.

## The Structure of a Software Project

A key aspect of conventional software design processes, in understanding how social contexts become structured, is the aspect of design described as "requirements analysis." The classic model of a software project comprises successive phases of requirements analysis, specification, implementation, test, deployment, and maintenance (Pressman 2008). Many variations exist in terminology and conduct, many of which include cycles of repetition and correction at various stages. Nevertheless all structured processes of software development commence with a concern to study and understand the needs of those who will use the final product, described as "requirements."

Requirements analysis is usually conducted in the field, in collaboration with the clients and end users of software. To echo the introduction to this volume, requirements analysis attempts to describe the context in which software will be used, with the aim of making explicit the presuppositions that underpin or inform design. It focuses on the ways in which the world is categorized, structured, and made sense of by users. It deals with complex social relations. It encompasses ideas, people, and things. It is more than descriptive process; it is a practice that entails abstraction, reflection, and production of second-order analysis. It requires sensitivity to the ways in which knowledge is recognized in different contexts. It is an inherently collaborative process. Indeed, as a mode of collaboration, it often becomes a means of critical engagement that intervenes in and transforms social contexts via the side effects of "business process reengineering."

Requirements analysis was a standard component of the software design process long before the first anthropologists were employed in technology companies. Yet the skills of the analyst (as such specialists were described) often resulted in a kind of fieldwork that suffered by comparison to academic anthropology. The timescales and resources devoted to requirements analysis are unusually small by ethnographic standards, requiring the invention of methodological shortcuts to gain as much knowledge as possible in a relatively short time. Furthermore, although computer software developers are highly reflexive, they and their software are primarily reflexive in a mathematical sense, making it difficult for field-worker analysts to appreciate the ways in which the software has interpreted their own contributions.

## Design

Of course, we can pay the same attention to anthropologists who claim to be doing design. The simple outward appearances of design work are often sufficient to identify professional status: Who held the pencil, operated the mouse and keyboard, stood at the workbench or drawing board? How many alternative concepts were worked up? What criteria were applied when evaluating the trade-offs? What languages (programming languages, schematics, plans) were used to engage in the sociotechnical collaboration? Who buys the product, and what choices do they have? Where is it used, and how often? How is it valued by users and by critics? Notwithstanding the epistemologies of technical and design work, these basic realities of day-to-day design are seldom applied to the work of anthropologists.

There is a vogue today for using the word "design" to refer to a professional stance or attitude to problem solving, especially in business contexts (e.g., Brown 2009; Martin 2009). This is accompanied in the academic community by a renewed interest in "design science"—the cognitive science agenda established by Herbert Simon (1969), in which computational models of structured problem solving are applied to larger-scale affairs. Although both of these tendencies are to some degree welcomed (with caution) by designers who agree that there are opportunities to apply design perspectives more widely, the combination of business enthusiasm and novel research funding programs has also led to some co-option of the word "design" as a buzzword, abandoning any reference to the actual work of designers. In cases where a design activity is not oriented toward any product, does not require any techniques developed via conventional design training, and does not involve any experienced design professionals, then designers and design researchers are probably justified in feeling uncomfortable.

Unfortunately, increasing engagement of anthropology with design has coincided with this vogue and has resulted in a confused public understanding of the ways in which design might be understood as a kind of applied anthropology. The design anthropology community, inasmuch as it can be thought of as such, can be seen to include both designers who treat "anthropology" as a cipher for any kind of engagement with a user during the requirements phase, and anthropologists who present their contributions to market research, public policy, or corporate strategy as representative of professional design work.

## The Structuring of Social Science

The tendency of business commentators to create buzzwords has been unhelpful to both design and anthropology. A surge in popularity of corporate anthropology as a business buzzword can be traced to an article published in *Business Week* (Nussbaum 2006). The central case study in this report was the activity of the People and Practices research group at Intel Corporation—the same company discussed in the chapter by Nafus in this volume. Journalistic coverage brought increased attention to the Ethnographic Praxis in Industry Conference, founded by Ken Anderson of Intel in 2005, and the anthrodesign mailing list founded by Natalie Hanson in 2002.

The first anthropologist hired by Intel (Genevieve Bell) joined the company in 1998, and in 1999 the *New York Times* reported on "Tony Salvador, an experimental psychologist at Intel, where he is called a design ethnographer," but anthropology had come to the attention of computer science far earlier. As reported by Lucy Suchman (forthcoming) in her history of anthropology at the Xerox Palo Alto Research Center (PARC), the first steps came twenty years earlier, in the summer of 1976, on the initiative of Jeff Rulifson, a computer scientist who managed the PARC Office Research Group. Rulifson first hired Eleanor Wynn, a graduate student in linguistic anthropology, followed by a team of three students who carried out fieldwork in Xerox branch sales offices the following summer, inaugurating a research tradition of workplace ethnography for the computing industry.

Rulifson sought assistance from anthropologists because he was dissatisfied with conventional business approaches to field research. Xerox PARC aimed to innovate by seeking alternative academic perspectives—in part, building on the work of Engelbart's human augmentation research project at the Stanford Research Institute, where working from a wide range of academic (and nonacademic) perspectives had become routine. However, business modes of design engagement with technology users had become systematized (e.g., through the increasingly conventional processes of requirements analysis) and thus associated with conventional products. In seeking to invent alternative kinds of product, it was necessary to create alternative, less conventional modes of engagement. Because the Xerox mission was fundamentally a scientific one, it was assumed that these modes of engagement should be grounded in a rigorous scientific discipline. Furthermore, because the Xerox researchers were creating software, the chosen discipline needed to be suited to the production of structured and systematic descriptions, capable of being translated into software.

It is notable that as the ethnographic research team at PARC grew, although the originally perceived requirement had been for anthropologists, some of the most influential contributors were trained in the sociological traditions of ethnomethodology and conversation analysis. Brigitte Jordan and Lucy Suchman developed techniques for ethnographic observation of the office context that relied on close structural analysis of video recordings, in an ethnomethodological style. Suchman's book *Plans and Situated Actions* (Suchman 1987) delivered a powerful critique to the established cognitive science focus of human–computer interaction research at Xerox. And when Xerox created a European research center (Xerox EuroPARC in Cambridge, England), Jordan and Suchman were dispatched to recruit ethnomethodologists from the sociology departments of British universities—in particular from Manchester, which provided two of the three EuroPARC directors.

The perspectives of ethnomethodology were assimilated into the software research establishment with particular urgency after Suchman's compelling criticism of the field. This soon led to demand among software designers who wished to apply the new perspective in their structured design processes. As already noted, in the software industry, a systematic approach to the design process is called not a "design method" but a "design methodology," probably in an attempt to appeal to external scientific authority when making contestable claims about social process. Unfortunately, this convention leads to an ambiguity in the case of "ethnomethodology," because the last thing that Garfinkel's work should be taken to represent is a prescriptive research methodology. On the contrary, introductions to ethnomethodology must always explain that it is an "ology" of "ethnomethods"—the study of the practices by which ordinary people (and scientists) accomplish social phenomena. Researchers at EuroPARC took pains to distinguish between the expectations of a design methodology and the commitments of Garfinkel's philosophy (Button and Dourish 1996).

Nevertheless it became increasingly urgent that the broader software industry be able to draw on the methodological innovations of the social scientists at EuroPARC. A typical textbook therefore elided Garfinkel's distinction, recommending that any software oriented toward a social outcome should apply "methodological ways in which ethnography may be utilized in the evaluation process and be married to design practices," leading to the definition of "ethnomethodologically-informed ethnography in the creative process of design" (Crabtree 2003). This insistence that ethnomethodology prescribes the right approach to design ethnography was most forcefully argued in a paper at the ACM Conference on

Computer-Human Interaction titled "Ethnography Considered Harmful" (Crabtree et al. 2009), which asserted that the only good design ethnography was ethnomethodologically informed ethnography—the "ethnomethodology" implied in the textbook.

Although there are few fields in which Garfinkel's work was so directly influential (science and technology studies being another), these events in advance of the popular corporate anthropology movement have significantly affected the way that ethnography is perceived among software designers. In particular, ethnography is associated with critique of cognitive accounts of the user, with opposition to certain technologies (those associated with artificial intelligence), with the extensive use of recording methods (especially video), and with a focus on the technologies of "collaborative work"—broadly, those in which multiple users interact simultaneously with the same system. The old styles of software industry fieldwork during the requirements analysis phase of design are discounted as being uninformed or even "harmful" in contrast to ethnomethodology.

This creates a further challenge when global corporations attempt to engage with phenomena in developing markets. Ethnomethodology is itself often accused of lacking political engagement: as expressed by Harris (mocking Garfinkel's terminology): "No amount of knowledge of 'competent natives" rules and codes can 'account for' phenomena such as poverty; underdevelopment; imperialism; the population explosion; minorities; ethnic and class conflict; exploitation, taxation, private property; pollution and degradation of the environment; the military-industrial complex; political repression; crime; urban blight; unemployment; or war" (Harris 1979, 285). The theoretical stances that underlie social critique are rejected by Garfinkel's ethnomethodology, just as firmly as it rejects cognitive accounts.

From the perspective of the software designer, seeking a structured account of a human context that will be commensurate with a design intervention, a social scientist appears likely to be a valuable collaborator. The relative merits of debates within social science are not easily evaluated by programmers and engineers.

**The Structure of Design as Research**

A significant factor contributing to the critique of software practices is a failure to recognize the distinctive character of design research. As should be clear from the earlier sections of this chapter, software design is routinely conducted in collaboration with social scientists or with others who have an understanding of the context of use that is complementary to the technical expertise of programmers. In the case of a software design research program, as opposed to routine software design, it is more important that these social science collaborators should have a rigorous and reflexive academic training—to an extent that their contribution to the design process is itself a piece of social science research.

Perhaps the most subtle implication of design research for those outside the profession is its relationship to commercial product design. Some forms of design research are purely observational, analytic, and critical—studying and commenting on the work of professional designers carrying out commercial design activity. However, design research in the technology field may also be practice-led or action research, in which actual design work is done as an integral component of the research process. However, for those not trained in design or familiar with design practice, this practice-led design research can subtly differ in further ways from other kinds of technology research, beyond the possibly unexpected integration of social scientists into the research team. In addition to being distinguished from observational and analytic studies of designers at work, it must also be distinguished from the professional use of fieldwork and user study techniques during the requirements phase of a professional design project, which is also (confusingly) called "design research" in some contexts.

The phrase "design research" can thus have three meanings: the academic study of professional designers at work, preparatory analysis within a commercial design project, and practice-led design projects conducted in an academic context. The outcomes of practice-led academic design research (sometimes called "research-through-design" in the technology design literature [Gaver 2012]) are, in addition to academic writing, a "product" of the design, which normally takes the form of a technology prototype—either a software or a hardware/software product. This prototype may be (indeed, often will be) placed within a context of use to evaluate results of the design process with regard to the design intention, or as a provocation or probe to elicit further understanding of user behavior, context of use, and manner of appropriation. In many respects, these research activities make it look as though a product has been made. However, as any professional designer understands very well, a prototype is not a product.

Prototypes are created in corporate research laboratories or advanced technology groups as a practical aid to exploring new technical concepts and business opportunities. This can be explained by analogy to practices in an older industry that is more familiar from twentieth-century popular culture: the relationship between the "concept cars" shown at motor shows and the manufacture of a new mass-produced model. Concept cars may have a working engine and drivetrain, but they might also be constructed from clay (traditionally), aluminum, or fiberglass, with wheels, transmission, and steering sufficient only to move them from a transport container to the display stand. Concept cars are handcrafted one-off display pieces, created as boundary objects between designers, marketers, managers, design commentators, and audiences.

When one of these reified concepts becomes the basis for a commercial product, it is understood that almost every aspect of the prototype design will be discarded. As a handcrafted object, it is not feasible for mass production. It will not have the right components to travel at speed, to carry loads, or probably to operate in rain, cold, or heat. Many of the most eye-catching features (such as transparent skin, video display screens, exotic materials) will have been simulated using elements that are too expensive or unreliable to appear in a real car. Furthermore, despite the expense of creating a one-off show car, the design process of an actual production car is far more expensive. The engineering effort involved in designing to a price, designing for manufacture, meeting environmental and safety regulations, and so on, is many, many times more expensive than the supposed public encounter of the design show.

All these observations can also be applied by analogy to understand the nature of a digital technology prototype. As a designer of both prototypes and commercial products, my informed estimate is that the engineering work involved in creating a commercial version of a completely new software product is at least ten times the cost of the original prototype development. This engineering cost does not include further costs of distribution, marketing, maintenance, training, and user support—all variable costs that increase in proportion to the size of the market and thus hopefully reflect increasing profit as they also increase. Nevertheless I estimate that for a modestly successful software product (one that makes some profit for its investors, some return for its inventors, and normal salaries for the employees of the company), the total costs would be one hundred to one thousand times the cost of the original prototype development.

These huge variations in scale should make us pause to ask whether the design work on which we comment in this volume represents actual products or research prototypes that may be minuscule in their economic and political significance. The prototypes constructed in collaboration with the ethnomethodologists at Xerox EuroPARC were used mainly by the PARC staff themselves, reflexively observing the consequences for their own social relations. Crabtree's textbook on ethnographic methodology is structured around a lengthy case study of a prototype catalog system deployed in his university library. In the present volume, many of the systems designed were created and tested for a specific group of users in a particular context and evaluated in that same context. All of them are excellent opportunities to conduct research-by-design, constructing prototypes that may challenge thinking about design processes.

## The Structure of Globalization: Inside Global Products

In contrast to the small-scale engagements of social scientists with technology researchers in a context of workplace ethnography and research-by-design, a very different view of anthropology is offered by experiences at Intel as reported by Nafus in this volume: "Firms hire anthropologists largely ... if they want to sell a new product in parts of the world they do not know about." However, whereas the earlier ethnographic work of Wynn, Suchman, and others at Xerox PARC was closely engaged with design research, Intel is not a design-led company (in the sense that its revenue depends primarily on products that it designs for end users).

As expressed in the branding slogan "Intel inside," Intel creates silicon components—indeed, was a founder of the microprocessor revolution. Unlike computer hardware companies (Apple, IBM, and Hewlett Packard) or software companies (Google and Microsoft), the main part of Intel's revenue still comes from the components that it sells to other companies. Where hardware and software companies rely critically on their expertise at designing products for users, Intel's greatest priority is to sell its products to the engineers who put them "inside." This places the company at a strategic disadvantage in applying ethnography to its business. It could attempt to study the contexts in which Intel components are chosen by the engineers at other companies—no doubt it does do this, but this work is perhaps too technically specialized to be performed by anthropologists (or at least too sensitive to be publicly reported by them).

Alternatively, Intel can attempt to second-guess its own clients by working in the field to identify the market trends to which those companies will be responding in the future. This market research at one remove is both challenging and unusually specific in the research questions that it raises. For example, from an engineering perspective, the single largest factor influencing the microprocessor requirements of a consumer device is the size of the screen. Most processing time is devoted to moving and calculating pixels. The number of pixels increases as a square function of the physical screen size and as a square function again of the screen resolution. These two factors—screen size and display resolution—are hugely significant factors for Intel's commercial future. In a discussion with Nafus, she reported a substantial programme of fieldwork that was largely concerned with predicting the future size of screens—an engineering consideration rather than an emergent ethnographic concern with screens as a contextually salient feature of the computer. Indeed, it seems likely that anthropologists working to a design brief from another company might find a completely different technical focus for their analysis, even if working with the same population.

It is unsurprising that the specificity of ethnographic description draws attention to the material aspects of digital technologies: the hardware of screens—their size and position—or the wires that connect them to networks, peripherals, or charging devices. Nevertheless the essence of software is its immateriality, not the relative size, cost, packaging, or power of particular pieces of hardware. Software designers are often frustrated by discourse that fails to engage with their actual work, as if a novelist faced a detailed critique of a book's cover and the cost of its binding that failed to mention the text. Although computer hardware continues rapidly to become cheaper, faster, and so on, the design of software is a surprisingly complex and slow-moving endeavor. Even at the surface level, the folders and trash bin of the Macintosh Finder prevailed for thirty years before starting to be supplanted, while the spreadsheet remains a design classic, with only cosmetic changes across the decades from mainframes to iPads.

**Design Research and Global Adoption**

When we consider the development life cycle of a globally successful product, as opposed to a laboratory design research prototype, we also find far greater diversity in relevant knowledges. The process of bringing a global product to market, including requirements analysis, specification, manufacture, pricing, and sales, is collaborative. Many specialists contribute to this venture. Many are educated in business, the arts, or the humanities. Many are local employees of multinational companies or specialist consultants in local market conditions. It is often the case that only a minority of those involved are technologists, in the sense that they are directly engaged in the technical design of the hardware, algorithms, and user interfaces that constitute the product.

In contrast, while a research prototype is a designed artifact, it is not a commercial product. It is constructed to test or challenge scientific understandings. In the case of a software prototype constructed for use by people in developing nations, the purpose of the prototype is to reflect on and challenge the prior understanding of software construction. However, it is important to recognize that such a research prototype is still not intended to be a product and is not designed as one. The adoption of a prototype by a notional market is not relevant to its purpose or evaluation, whether that market is among colonial powers or indigenous people. This can be a source of tension and confusion when computer scientists conduct research in the field.

The research community that has a special interest in creating and deploying ICT for developing regions is generally known as ICT4D (ICT [for] Development). I am not an ICT4D researcher, so I address this topic with caution; interested readers should consult more authoritative sources (e.g., Heeks 2010). Nevertheless my analysis is based on an extended period of engagement, during which I directed a national funding program closely linked to ICT4D research. Bridging the Global Digital Divides (BGDD) was sponsored by the Engineering and Physical Sciences Research Council (EPSRC), the main funding body for technology research in the United Kingdom. I helped conceive the program, managed the process of recruiting interdisciplinary teams and commissioning specific research projects, and acted as a strategic consultant and adviser to the program throughout its operation.[1]

I do not have space here to describe the BGDD projects in any detail, but two examples indicate the flavor of the design research carried out. The Fair Tracing project worked with fair trade marketing organizations and independent wine producers in Chile to establish closer relationships of trust, via the Internet, between people who produce fair trade products and people who consume them. The StoryBank project worked with a community radio station in rural India to support local people with low literacy in collecting and sharing narratives of village life. Rather than specialized equipment, mobile phones were used to create and edit effective illustrated stories. These stories became both a local community archive and a resource for design students in the United Kingdom to conceive products that took account of other life experiences.

I take BGDD as a case study to illustrate the opportunities for design research in the context of science and technology policy in the United Kingdom. The choice of individual project objectives represented a fine balance between goals that were technically achievable within the scope of the program (so involved little technical uncertainty or adventure) while also demonstrating the application of relatively advanced technologies that were not yet familiar outside technology research contexts. The program's fundamental concern was to explore the ways in which advanced technologies can be shaped by encounters with other contexts, at a stage in their evolution where the associated design practices and assumptions are still malleable. This technical focus contrasts with that of ICT4D research, whose main concerns relate to public policy and the management of technology deployment and use. ICT4D integrates concerns from the fields of development studies, management science, information systems, and the broader social sciences—geography, anthropology, sociology, and economics. This volume itself might well be considered a contribution to the mainstream literature in ICT4D. However, because ICT4D is based in the social sciences, UK research in ICT4D would normally be funded by the Economic and Social Research Council, not the EPSRC.

The implication to be taken from EPSRC funding was that BGDD was neither a mainstream ICT4D research program nor a social science program. This confused many observers, especially considering that many of the researchers employed as members of the project teams were geographers, economists, and educationalists, outside the normal funding remit of the sponsor. On one hand, some senior members of the EPSRC research community considered that the program represented a misuse of public funds. On the other, members of the ICT4D community questioned the value of the research as a contribution to the discipline because of its focus on novel technology that might better be replaced by existing products. Such critiques are not unique to the BGDD program. ICT4D authority Geoff Walsham has commented that systems designed for use in the developing world are almost never adopted for long-term use in the intended market. The same observation could certainly be made with regard to the systems designed during BGDD.

However, the technical outcomes of the BGDD projects were design prototypes, not designed products. One would not expect prototypes to be adopted for long-term use in any market. If we look instead to ordinary products, rather than research prototypes, we see that many forms of software are already adopted and used in developing countries, including mobile phones, GPS receivers, automobile engine management systems, Web sites, search engines, monetary payment networks, and so on. Many of these ICTs rely primarily on the deployment of global technical standards (the number of bits in an IP address, the frequency distribution of a carrier signal, the HTML tags defining a Web page), with no special design attention to developing nations or indigenous users.

The engineering and business factors that distinguish research prototypes from standardized products are the single greatest determinant of the relationship between practice-led design research oriented toward users in the developing world and the concerns of the ICT4D community. The development of a technology prototype, often in collaboration with social scientists, is carried out with a budget that is one-hundredth to one-thousandth of the budget that would be required to achieve wide-scale adoption of a similar product. Many academic technology specialists, let alone social science academics, have never worked in a professional design context and hence have little awareness of these factors—especially considering the "motor show" phenomenon, that concept demonstrator prototypes often look like (and are made to look like) real, if futuristic, products.

So what is the proper ground for critically assessing programs like BGDD, if it is not possible that the products created could ever be adopted? The technologists, design researchers, and sponsors engaged in BGDD were all thoroughly aware of these economic dynamics; they apply to all technology research, and EPSRC would never expect that a prototype created in a publicly funded research project would be adopted as a product unless followed by major business investment. Some social science collaborators within the program had not worked as members of design teams before (although many had), so it was necessary to ensure that their expectations were realistic. But most challengingly, the program's creators had decided, as a matter of principle, that all the funded projects would incorporate extensive engagement with local stakeholders, field visits by as many of the team as possible (including technical specialists), and extensive partnerships with local organizations, businesses, universities, and NGOs. For these local collaborators, it was essential that they understand the nature of their participation as a research collaboration, not as an opportunity to acquire working advanced technology products—although almost every aspect of the research made it appear to them as though technology products were being created for their use.

### The Knowledge Systems of Subversive Design

The case studies described in the present volume do address significant epistemological and ontological differences and tensions. However, these are not matters of design; more often than not, they are differences between the knowledge systems of Euro-American science on one hand and indigenous people on the other. To the extent that design has been carried out at all, it simply reflects the encounter between the varying stakeholders. A competent designer, especially a competent design ethnographer, always manages processes of engagement with stakeholder communities, specifies software that is informed by the ontological commitments of the various users, and reflexively constructs a product with the potential to act as a boundary object or trading zone between them. The designers of the systems described here may have been more or less competent—but neither they nor their design processes have been the focus of study.

Anthropologists play another important role in design research—not simply as commentators but as participants and collaborators. I hope that anthropologists continue to engage directly with computer scientists and design researchers who recognize the value of a "critical technical practice" (Agre 1997), rather than disengaging from the design process and commenting only on the resulting products. Technology design research offers many opportunities for subversion, conversion, and development—including the opportunity to convert the process of development itself, in which case design research should be the greatest subversion of all.

**Note**

1. I did not receive research funding from the BGDD program myself. I managed the commissioning phase as a part-time secondment from my regular university work.

**References**

Agre, Philip E. 1997. *Computation and Human Experience*. Cambridge: Cambridge University Press.

Blackwell, A. F. 2013. The craft of design conversation. In *Software Designers in Action: A Human-Centric Look at Design Work*, ed. A. van der Hoek and M. Petre. Abingdon: Chapman and Hall/CRC Press.

Brown, T. 2009. *Change by Design: How Design Thinking Transforms Organizations and Inspires Innovation*. New York: HarperCollins.

Button, G., and P. Dourish. 1996. Technomethodology: Paradoxes and possibilities. In *Proceedings of the 1996 CHI Conference on Human Factors and Computing*. Austin, TX: ACM Press.

Crabtree, A. 2003. *Designing Collaborative Systems: A Practical Guide to Ethnography*. London: Springer.

Crabtree, A., T. Rodden, P. Tolmie, and G. Button. 2009. Ethnography considered harmful. In *Proceedings of the 27th International Conference on Human Factors in Computing Systems* (CHI '09), 879–888. Austin, TX: ACM Press.

Crookes, D. 2012. Computing experts welcome ICT shake-up. *Independent*, January 11, 2012. http://www.independent.co.uk/life-style/gadgets-and-tech/news/computing-experts-welcome-ict-shakeup-6288049.html.

Gaver, W. 2012. What should we expect from research through design? In *Proceedings of the 2012 CHI Conference on Human Factors and Computing*. New York: ACM Press.

Harris, M. A. 1979. *Cultural Materialism⊠: The Struggle for a Science of Culture*. Walnut Creek, CA: AltaMira Press.

Heeks, R. 2010. Development 2.0: The IT-enabled transformation of international development. *Communications of the ACM* 53 (4): 22–24.

Martin, R. 2009. *The Design of Business: Why Design Thinking Is the Next Competitive Advantage*. Boston: Harvard Business School Press.

Nussbaum, B. 2006. Ethnography is the new core competence. *Business Week*, June 18, 2006. http://www.businessweek.com/stories/2006-06-18/ethnography-is-the-new-core-competence.

Pressman, R. S. 2008. *Software Engineering*. Columbus: McGraw-Hill.

Simon, H. A. 1969. *The Sciences of the Artificial*. Cambridge, MA: MIT Press.

Suchman, L. 1987. *Plans and Situated Actions: The Problem of Human-Machine Communication*. New York: Cambridge University Press.

Suchman, L. Forthcoming. Consuming anthropology. In *Interdisciplinarity: Reconfigurations of the Social and Natural Sciences*, ed. A. Barry and G. Born. London: Routledge. http://www.lancs.ac.uk/fass/doc_library/sociology/Suchman_consuming_anthroplogy.pdf.

World Wide Web Consortium (W3C). 2012. Ontology for Media Resources 1.0. http://www.w3.org/TR/mediaont-10.

Woolford, K., A. F. Blackwell, S. J. Norman, and C. Chevalier. 2010. Crafting a critical technical practice. *Leonardo* 43 (2): 202–203.

# 10 Design for X: Prediction and the Embeddedness (or Not) of Research in Technology Production

Dawn Nafus

Consumption has been theorized as a process of appropriation (Miller 1997) where consumers adjust, reframe, hack, or otherwise infuse their own meanings and intentions into the commodities they buy and use. Information and communications technologies (ICTs) are no exception (Horst and Miller 2006; Burrell 2011). Designers of ICTs are increasingly recognizing consumer appropriation in various ways. One way is to hire anthropologists and user experience researchers to conduct what is called user-centered design. This enables firms to better understand and anticipate the ways consumers might appropriate their products. Although fully predicting those uses is impossible, the premise of user-centered design is to narrow the social distance between users and designers so that products stand a better chance of being valued on the market. While user-centered design purports to extend consumers' agency by involving consumers in the design process, its emphasis on research also slips between an anticipation, where one simply makes better bets, and a more problematic rendering of people into predictable, and one might daresay docile, beings. This tension calls into question the relative agencies involved in appropriations of technology. Do the feedback loops of user-centered design ultimately leave consumers tamed by products' designers, or do their appropriations somehow not count in the world of product development? Who is appropriating whom?

This is not a question we can answer directly. We can, however, take a step back and problematize prediction as itself a socially situated process that creates various forms of agency for different kinds of actors. Here I reflect on acts of prediction at a large multinational technology firm, Intel, not just in its user-centered design functions but also in strategic planning and other market functions. I discuss my role as an anthropologist in industry, where I made my own attempts at finding a productive way of destabilizing what I saw as certain negative aspects of predictability while still facilitating economic exchange. This brought me rather unsuspectingly into renewed scholarly debates on the usefulness of "embeddedness" as an anthropological concept that have arisen in response to Michel Callon's (1998) work. What I find most productive about this debate is the sense that connection and disconnection between economic participants might not be an either/or proposition. As people negotiate with whom and what they are embedded, and how and when they are quits, embeddedness is a moving target. In this chapter, I reflect on how predictions of what users will do has a similar both/and quality in the way that it too simultaneously creates both connections and disconnections between firms and consumers. Firms both organize themselves around what users will do and, simultaneously, consider themselves quits with consumers after having exchanged products, fulfilled warranty obligations, and so on. Trafficking in knowledge about what users will do creates connections between people and externalities in the very same act. In the language of science-studies-inflected economic sociology (Callon 1998), predictions create frames and overflows as knowledge is shuttled between a firm and the consumers of its products.

I suggest that the constant embeddings and disembeddings of various knowledges are not just useful theoretical tools but do work on the self. Overflow—Callon's term for the excesses of social connections that burst the tidy frames that neoclassical economics would have people confine themselves to—is usually treated as a theoretical concept to parse an economic situation. Here I approach overflow as an embodied experience. Where there are multiple entities in which one might be embedded, and these entities largely see each other as external to one another and yet related in some way, these overflows are at best challenging to negotiate. They force certain acknowledgments of one's own agencies and limitations, and how those agencies extend through objects and measurements as well as people. For me, these experiences of always exceeding the current frame have been at times exhilarating, at times frightening and disheartening, and frequently all three at the same time. Overflows are at the heart of the daily experience of doing social science in a nonacademic context. Embedding and disembedding, especially when the two result from a single act, are processes as laden with affect as they are rational calculations to retain influence within the firm. There are serious personal costs even after naive preoccupations with having "sold out" have faded.

I ground my thoughts in a project that attempted to create space for expanded consumer agency not by designing a particular product but by challenging in a more general way what a technology firm needs to be able to predict about its consumers to have something to exchange. The project

# Design for X

started life as an interest in the need for prediction in the context of problems of planned obsolescence but ultimately focused on what needs to be made predictable about consumers in the so-called emerging markets. I first reflect on the social relations within technology production that produce the need to predict uses of technologies, starting with a short account of how I went from a concern with planned obsolescence to emerging markets. I then explain the details of the research project that was meant both to work within this set of relations reliant on prediction and to thwart some of its negative consequences. Finally, I reflect on the forms of agency I found myself caught up in, and the limitations of being able to predict the effects of my own objects as well as the firm's.

## From Reuse to Institutional Systems of Production

There has been a growing recognition within industrial design that leaving greater scope for redesign and repurposing makes for greater environmental sustainability (McDonough and Braungart 2002). An example of this recognition is the packaging for POM, a beverage made from pomegranate juice. POM is sold in glass containers that, once the top is removed, become a reusable drinking glass. While most bottles can perform this function if pressed into service, the design facilitates it more pleasurably and inspires people to reuse the container in this way. Outside of professional design, there is enthusiasm for reuse and repurposing. Publications such as *Make* and *ReadyMade* explicitly celebrate everyday acts of consumer appropriation—old window shutters remade into a garden bench, shipping tubes made into a DVD holder—as acts of resistance to mass manufacture and its environmental damage. In urban, fashionable quarters of the United States, craft, hacking, and self-production of all kinds are seen in many quarters as important ways out of the environmental costs associated with mass manufacture (see Schor 2010).

Combined, these new sensibilities among both designers and consumers represent an important shift in how the relationships between design and consumption are constituted. There is a clear shift toward a greater acknowledgment of consumer agency. This acknowledgment goes beyond consumers' ability to choose among a variety of goods in a market and extends to their ability to reshape the material world. Yet there is a sense in which the loop is not helpfully being closed. User-centered design still struggles to fully engage with the prospect of reuse, and for good reason. Designers might build products in such a way as to be more easily taken apart for remanufacture when the conditions of remanufacture are known, but design alone cannot ensure a product's ultimate trajectory. A person might recycle it, might follow the suggested reuse, or might do something totally different. She might just throw it out!

My preoccupation with how large firms might engage differently with consumers' agency started out as an awareness of these transformations. How might a large technology firm come to value its consumers' DIY sensibility in such a way that its products are less tied to planned obsolescence? Traditional business models in the ICT industry are still tied to planned obsolescence, and this would mean not just redesigning products but also the points in the exchange at which money is made. One possible direction was to muster ethnographic evidence to mount an argument for Intel to develop a particular PC, or part of a PC, that eased the process of repurposing. It would remain in physical circulation for quite some time and be tied to a service business where Intel could make its money on the nonmaterial aspects of the exchange. Alternatively, perhaps I could argue that computing might be applied in some way to the processes of reusing and reshaping other objects, increasing demand for computers while decreasing demand for other commodities. That is, I could construct a view of a market that would give the firm economic reason to expand its commitments to repurposeable objects.

Designing computers for appropriation is a complicated social and technical problem that others have thought about (Dourish 2003), but it reveals underlying power dynamics at stake in user-centered design. Being able to predict what someone is going to do is fundamentally a power-laden act. A merely predictable consumer is a docile one. Consumer docility is a fiction, of course, but a dangerous one when held by elites who can choose what will and will not be produced. A wildly unpredictable consumer, on the other hand, is one that firms easily free themselves from the obligation to comprehend. The social shaping of technology literature (see, e.g., MacKenzie and Wajcman 1999) shows that when actual users are ignored, technology producers make what they believe to be a general-purpose product that on the surface appears neutral but always, in the end, is inflected by the sensibilities of technical elites. The market then can serve as a way to reproduce those sensibilities, against which consumers who do not fit the mold must work even harder as they appropriate technologies for their own purposes.

When I thought about the problem of sustainability, one conclusion I came to was that encouraging the development of a particular kind of computer, optimized for maximal reuse, was important, but it was only one of a number of efforts that needed to be made if the consumer agencies involved in repurposing were to be taken seriously. When design for reuse became the goal, it became clear that technology producers are too often organized around either predictable consumers or irrelevant consumers, and only rarely agential consumers. I discuss some of this organization later. For now, it suffices to say that I came to see that consumer agency was being avoided for much broader reasons not necessarily tied to an indifference to sustainability or the lack of clear demand for reusable devices. Together with other social science colleagues at Intel, I took this as an opportunity to look instead to reshaping decision-making processes. How might we get beyond the modalities of approaching consumers as a mechanically predictable target market or ignoring them in the service of developing highly suspect "general-purpose" technologies?

In parallel, similar issues arose in the context of assessing consumers' behavior and values in the so-called emerging markets (that is, countries with middle income levels). This was territory with which I was much more intimately familiar, but by identifying the systemic problems associated with sustainability, I came to see it in a different light. Frequently I am asked what is the best design for emerging markets. I found the only honest answer was that I did not know. There is no such preexisting cultural grouping outside of firms that need to make claims about who people are and what they want from technologies. As a participant in user-centered design for emerging markets, I was finding that no matter how compellingly I advocated for, say, the only slightly more specific needs of first-time computer buyers in Russia or China, it still did not necessarily mean that their desires and sensibilities were addressed in ways that I believed would be meaningful to them. More importantly, I saw little evidence that the Russian or Chinese people I cared about were necessarily being approached as fully agential, socially situated beings. Whatever I said ended up becoming an abstract caricature in the service of making "emerging markets" imaginable as a large-scale market. Even if we ended up with a good product likely to be valued, the process of getting there felt organized to discourage any sense of empathy and connection with others by abstracting people into addressable markets. This was true even though my colleagues at Intel on the whole genuinely do want to create meaningful connections with the people who buy their products, and do not lack human capacities for curiosity or empathy.

I had originally taken this traffic in caricatures to be a personal failing owing to my own lack of political skills. The problem of consumer reuse showed me that I was running up against a broader social constitution of consumer agency and misrecognizing it for my inability to use my own agency. What sorts of uncertainties and consumer agencies can and cannot be accommodated or supported within a multinational technology firm like Intel thus became an empirical question. The problem of sustainability revealed to me, more starkly than I had previously been in a position to see, that forms of consumer agency are socially constituted not just in the act of consumption but also within technology production itself.

## The Social Organization of Prediction

The prospect of consumers as agential beings challenges, in a fundamental way, widespread notions of how businesses relate to markets. These notions are not just abstract beliefs but shape how businesses are organized formally and informally. This shaping, and the ways in which a full acknowledgment of consumer agency challenges it, is what brings my work into the recent debates in economic sociology and anthropology. Prompted by Michel Callon's *The Laws of the Market* (1998), these discussions are to some extent a refinement of the much older formalist-substantivist debate within economic anthropology (for a review, see Plattner 1989). This earlier debate explored whether economic relations can be thought of as embedded in and subsumed by other relations, such as kinship (the substantivist approach), and the senses in which economic action can be analyzed as a matter of abstract calculation on the part of individuals (the formalist approach). The formalists did not assume calculation always to be a function of maximizing monetary profit, but they did squarely emphasize individual calculations to gain whatever is locally valued (prestige, etc.).

The twist that Callon provides is the notion that neoclassical economics, with its presumption that rational, self-interested actors are always asocially, autonomously calculating their gains, has itself become embedded into everyday life. It is a frame with which actors themselves understand the market and act in that market, though such a framing posits that people are disentangled from one another. When formal economics became culture, its abstractions became concrete reality because enough people performed them in the course of everyday actions. Miller (2005) charges that despite the undeniable embedding of abstract economics in real market relations, privileging it nevertheless plays into problematic processes of virtualization. The economists' formal models, for instance, have had the greatest uptake in the financial markets, and noneconomic reasoning is much more readily apparent in the process of buying and selling real objects like cars, rather than virtual objects like derivatives that only years later would prove so massively destructive.

Within this discussion, Slater (2002) helpfully suggests that we return to the notion of frames, which Callon took from Goffman to think through how people can think of the economic as if it were a separate domain where actors are quits after transacting, when they can also readily see that there is more going on. Framing things as "economic" cuts off all the entangling relations also in play. Slater uses this always present relationship between frame and overflows to reminds us that the disentanglements that framings perform are only momentary. He returns to Miller's example of a woman purchasing a car. There is in fact a wide range of cultural and social entanglements that bring the woman to the car dealership, render the car meaningful and consumable, and so on, but at the moment of transaction they are both in fact quits, because they both assume that the transaction itself does not incur additional obligations to each other. Other relations may be going on around it, even in possible tension with this transaction, but this market transaction is indeed a distinct form that summons the notion of self-interested disembedding. Slater provides a complicating point of view that suggests more is going on than the performances of neoclassical economics. In fact, Callon himself observed early on (1998) that for some, the entangling relations are problematic excesses to be managed, and for others they are the social embeddedness to be celebrated.

Frames and their overflows help us understand how technology markets are organized. The organization of many businesses, Intel included, is premised on the imagined ability to draw an arrow between what people do and believe—that is, a target market—and a product that the business hopes that market will value. These imagined connections shape internal organizational structures, where and how manufacturing takes place, what kinds of research and development investments are made, and so on. However significant this arrow is to the functioning of the business, it always points outside the firm. Outside there are people with whom it can be quits after transactions occur. By drawing this arrow and pointing it outside, firms create their own externalities (Strathern 2002b): a group of people who are outside themselves, with whom they incur no further obligations, and whose concerns can therefore never be fully internalized, but also never fully disentangled, either.

Designers, social scientists, market researchers, product planners, advertisers, and strategists all sit between markets and production and have highly particular, prescribed roles in pointing these arrows. In this, my role as an anthropologist in industry is not unique in the slightest. Callon has much to say about mediators between firm and market. He advocates that we should "refuse the reduction of the economy to a tragic face-to-face between production and consumption. ... Markets include a large number of actors and *agencements* that multiply the gaps, differences, and disagreements" (2005, 15). In his view, violating the frame of the strictly economic by negotiating these social overflows does not tidily produce consensus, as in the image of the anthropologist as a kind of translator and smoother of relations, but rather multiplies differences and gaps in understanding. While Callon celebrates these gaps by calling them the opposite of tragic, later I will show why I am far less triumphant about them.

The frame of the firm as external to its markets acts as a stabilizing mechanism that enables people inside it to forget that gaps, differences, and disagreements do proliferate by drawing these arrows. For example, at the time of writing there was regular disagreement about whether the use of a small-screened computer encroaches on the market for a large-screened one. Whether use of the two products is deemed to be complementary or encroaching is not at all merely an "external" matter of what users will do with products. The outcome of the debate can trigger a costly and risky institutional reorganization and reallocation of resources. Yet despite this precariousness, the language that people use to describe the relationship between technology users and various people within the firm is suspiciously stable. Indeed, at Intel entire job descriptions are structured around whether people are closer to the end user or farther from them. Hallway conversations are peppered with talk of "Is he customer facing?" or "You need some use case people on your team—you are too far from the consumer." It is as if these were unproblematic matters of fact, while the next reorganization remains a matter of deep, persistent concern. So stable are the connections between firm and market in everyday talk that it is some people's job to build a "go-to-market engine," a system of intrafirm relations that bring a computer processor into computers, and those computers into retail or other environments for purchasing. The metaphor, of course, suggests that the firm does not intend to fire up the engine only once—it is a repeated, stable assemblage that people believe holds little capacity for differences and disagreements. Yet one can readily see the unmachinelike ways in which these assemblages really work in the equally prolific calls for greater teamwork, cross-organization transparency, and the inevitable jostling to establish position before or after a reorganization—activity that would make little sense if the machine indeed worked as a machine, and differences were not proliferating.

Firms hire anthropologists largely because if they want to sell a new product in parts of the world they do not know about, it helps to have someone who can anticipate the likely assumptions people will make about

# Design for X

the product. This is particularly true of the technology industry, which largely imagines technology to have social effects as if standing away from the social at arm's length, rather than being already included in it (Strathern 2002a). Technology exacerbates a sense of social distance beyond what might be produced by the externalities of market relations alone. Technology users constitute distant exotica for technology companies dominated by engineers, requiring a pith-helmet-wearing explorer to reveal them (Nafus and Anderson 2006). In this way, anthropological research in the technology industry started as a kind of salvage operation to recover the social embeddedness that engineers find difficult to imagine. There is a parallel here. While far less pervasive, the urge to rehabilitate social relations through anthropology is also embedded in and performed within the marketplace, in ways roughly analogous to Callon's performative neoclassical economics. To the extent that market actors adopt what anthropologists know, those performances of anthropological knowledge also have consequences.

There is an assumption in industry that anthropologists work within a stable feedback loop where any redesign by users that they discover might be incorporated back into the next generation of production. In my view, this practice is closely linked with practices of risk reduction. Intel remains one of the largest and longest-standing social science research practices in industry. Not coincidentally, it has extremely long production cycles compared to other firms in the industry, and therefore particularly high intolerance for risk. The goal is to help the firm anticipate long-term social transformations that affect its business, given that its production cycles are equally long-term. While there is an understanding that taking computer users seriously ultimately produces a better product, it is an understanding rooted in the need to reduce the risk that a product will fail given the enormous investments required to retool factories and orchestrate "go-to-market engines." Put differently, for firms like Intel, what matters about consumers is only what can be anticipated with some confidence. Confidence requires knowledge reduced to an expressed need for a particular product feature that can be put into a product manufacturing schedule with equal confidence that it will be bought.

This poses a problem for us interlocutors: how to contribute in such a way that human concerns remain a part of the conversation and do not get absented by a presumed neutrality of technology, but also where those concerns do not get reduced to the need to "design for X," where X is an overprescribed sociotechnical configuration: the Chinese want this, the Russians want that, and what they happen to have in common gives us our emerging markets when totaled up. Too often, X can be anyone and no one, all at the same time. At a global company that needs to sell more or less standard products right round the world, X scales and rescales with peculiar plasticity. Sometimes Russians plus Chinese plus Brazilians plus Indians equals "emerging markets," while sometimes Chinese plus Germans plus Americans become the "mobility market" through some selected commonality. Sometimes X is one man in Kenya using his mobile phone to make a payment to his rural mother. In some circumstances, he can be made to stand for the "netbooks market."

There are limits to the ability of ethnography from any one of these places to change these scaling practices, which are, in Callon-like ways, constitutive of market activity. At a certain point, the otherwise confidence-inducing reality of ethnography does damage to the need to imagine a market that does not yet exist. It is a *future* market, necessarily a simulacrum of what is not yet there. X cannot be so narrowly specific as to not be conceivable as a viable market, but even the most fast-and-loose strategic planner would not want X to be so open to negotiation that it can be made to fit and legitimate whatever fantasies are currently in favor. The simulacrum of a market that does not yet exist needs social friction to generate traction, but not so much that it breaks the imagined scalability that convinces actors to act at all.

**Creating a Boundary Object: The TDI**

Under these conditions, a small group of anthropologists within Intel sought to reshape our work in a way that loosened the need for predictability. Prediction can never go away entirely, nor is that desirable from a business or research perspective, but perhaps we could evade the plastic, caricatured constructions that made scalability so distant from human concerns. We approached research as a designable object—not in the ordinary sense of having a research design, but as something that acted in more objectlike ways. It would be reasonable to say that we created a "boundary object" (Starr and Griesemer 1989) to buffer between our own expertise and the actions of the firm. A boundary object is an object that has different meanings in different social worlds, but aspects of it are common to both such that it facilitates relations without attempting to achieve agreement. We had learned that agreement was not working; ethnography was being subsumed in processes of making markets that scaled. Instead the task was to design a boundary object that mobilized key aspects of anthropology in particular ways.

In the previous description of how markets are created in the imaginary of technology production, prediction was necessary both because long time frames are associated with product development and because products had to be sold across social contexts, not simply for one. Anthropology is no stranger to questions of scale (Strathern 2004). How anthropology addresses scale, and how strategic planning addresses scale, are related yet incommensurate. To oversimplify, anthropology takes something quite small, quite particularistic, and says something much more general on the basis of it. The planners and engineers that I work with often struggle to get beyond the particularistic description, which appears more "real" than the theoretical work. Elsewhere ken anderson and I have called this the "real" problem (Nafus and anderson 2006). It makes a certain sense that someone who has to manage a product line that gets shipped all around the world would not want to spend an hour of his time talking about kinship practices in urban Russia, and what technologies could do to better support that, when it seems highly unlikely that kinship practices could possibly be about more that "just" urban Russia.

Traditional market research, on the other hand, has excellent ways of creating pliable scales. Consistent survey protocols with comparable quantitative results may not be as "real" as ethnography, but they are more consistent with how product planners imagine a market as a quantity of actors who we can predict with some confidence will buy a certain product. This being the case, we switched strategies away from dazzling product planners with the specificities of the field. Instead we asked what kind of generality would allow them to make a different reading of the specificities. We knew there were unsaid generalities that made some scalings more easily interpretable than others. If Russians plus Chinese plus Brazilians plus Indians could equal "emerging markets," it was not because research had determined that technology users in those places had anything in common. It was because product planners considered places that they already understood to be "mature," and the rest "emerging."

Distinctions between mature and emerging markets are the private-sector equivalent of developed and less developed. The problems with this distinction are, of course, well known within anthropology. It casts places that are unfamiliar to white people as emerging from some heart of darkness, infantile until they are shaped into markets intelligible to multinational companies and multilateral development organizations. The trope renders social change into a loaded story about progress, such that it becomes impossible to question the necessity of development, only the specific means by which a country, like a child, might reach maturity (Ferguson

1994). In these ways, a firm's need to understand "emerging markets" in the first place is structured by distinctions between the West and the rest, in which clear cutoff points exist at key levels of gross domestic product (GDP). GDP-based cutoff points are neoclassical economics' abstractions, as embedded in the daily practice of transacting goods as the wider sense of social superiority from which they come. They are performative and give shape to markets. This powerful logic meant that every time we researchers were able to table our regional scholarship, the ethnography was seen with this lens. Thus kinship practices in, say, Russia, were really only minor indices that suggested either maturation or a lack of it. The dominant distinction was "poor or not poor," regardless of the ethnographic facts.

The people involved in these business processes are not fooled. They know the absurdity of talking about Shanghai as part of an emerging market, though the calculations that enact markets would suggest otherwise. Pointing this out is not new knowledge, but there are ways that measurements of markets perpetuate the absurdity. Take, for example, the IBM/World Bank e-readiness index. The index compares countries' level of technology "maturity" by combining technology adoption rates with other assumed markers of maturity such as GDP growth, low tariffs, government transparency, literacy, and so forth. The e-readiness index finds, unsurprisingly, that the countries that already have the most technology are the ones most ready for it.

Indexing has the advantage of providing a single criterion for selecting markets, either for the purposes of developing a new product or prioritizing countries in which to build marketing and other institutional capacities. Indices do not dictate how Intel approaches emerging markets, but they do build the case that mature and emerging markets are separate and distinct from each other, regardless of the technology under consideration, and a high GDP is a marker for which market is ready for which technology. They build the case in part because they put in economic terms a preexisting cultural model, but also because they are disembedded from the firm. If one can point to outside parties such as the World Bank, it is easier to deal with a slighted manager, charged with sales in a particular country, who wants to know why his country has not yet been deemed mature and is therefore not receiving the appropriate resources (or vice versa). In this way, pointing to the disembeddedness of knowledge, regardless of how culturally situated it actually is, can help embed ones' own perspective into economic action.

My colleague ken anderson and I teamed up with a quantitative sociologist, Phil Howard, and produced our own index, which we called the Technology Distribution Index (TDI). Our index retains the logic of ranking but measures the amount of technology adoption *relative* to a country's income level. It measures a country's share of the global stock of a given technology relative to its share of global wealth (for a full explanation, see Howard et al. 2009). This means it is possible to say which countries adopt technologies at rates on par with their economic power, and which ones adopt at higher or lower rates. This opens up different kinds of comparisons between India and China, or Kenya and the United States, because the dominant comparison (wealth) has been mathematically leveled. It also surfaces places with surprising levels of technology adoption. Instead of the usual OECD suspects dominating the top of the index as they do in e-readiness, at the top of our list are places like Estonia, South Korea, Morocco, and Turkey. The measure is not necessarily a singular one. For some purposes, we aggregated an average for all ICTs we measured (PCs, mobile phones, Internet users, broadband subscribers, broadband bandwidth), but for some we aggregated averages across the technologies that matter for the particular problem at hand. For example, if product planners are looking to develop a highly mobile device, a high score in PC ownership says less about the market readiness for the new device than a high score in mobile phones if adopting the device indeed requires skills of mobile phone use. We spent time helping planners to disaggregate our index and figure out which measurement to use.

The index is a technology of simultaneous strangeness and familiarity. Because Intel also uses other indices, lists of countries to target for this or that product already have a circuit through which they circulate. Yet the strangeness of the list, mixing poor countries with rich ones, frustrates the urge to predict what market is ready for what technology based on preexisting logics. We worked with a designer to create a world map that redrew the imagined geography in surprising ways (fig. 10.1). China and Brazil are still "large," as is Finland, but so is Kenya, Vietnam, and much of eastern Europe. India is not, reflecting its mediocre technology adoption relative to its income.

From a business standpoint, the index scales up to a level where trade-offs can be made between investing in this market or that, thus satisfying the need for generality. Yet it leaves open the possibility of generating contextualizations by marking out the relation between technology and income without suggesting what is generating that relation in a particular country. The map is a visual rhetoric of completeness that uses the question "what's behind these numbers?" to generate a better discussion of what social and cultural phenomena besides disposable income make technologies valuable to people. The TDI makes it harder to imagine that the poor

214  Dawn Nafus

**Rate of Acceleration/Deceleration of Technology Adoption, by Country**

#### Figure 10.1

Technology adoption relative to wealth. Although this graphic was originally designed for color (full color image a: http://www.wired.com/images_blogs/wiredscience/files/tmi_2007_global_map_13.pdf), here we can see that the countries that adopt technologies at high rates relative to wealth do not follow a geographic distribution traditionally associated with levels of development. Source: calculations by Howard et al. (2009) based on World Bank and ITU data. Graphic design by Steven Marsh.

are buying technologies only because they are now less poor. This enabled us to have a conversation with people at Intel about what these places might have in common if not development.

We were clear that this is also a panoptical rendering of the world. It is an entire map, seemingly complete, and on the surface a view from nowhere and everywhere at the same time. Yet this disembedded "god trick" (Haraway 1991) is precisely what situates it within an industry that explicitly eschews designing technologies for a particular place. In this way, it acts as Callon would predict. Like the calculations of neoclassical economics, the TDI is an object that pretends universality, and that pretense frames a socially situated marketplace. By seeming to be disembedded from human context, it speaks to market actors in a way that an argument for designing for the cultural values of this or that place could never do. Yet it also is a deliberate attempt to renegotiate the overflows from that market framing. It makes local difference visible by creating an unanticipated overall pattern, which gives product planners and strategists a different way of thinking about how to act on those differences, rather than rounding them all up to the mess of "emergingness." In this way, it embeds cultural differences into strategy questions without specifying what those differences are, exactly, or reducing them to stereotypes. Thus what X will do with what technologies can be somewhat anticipatable, without claiming full predictability based on a story that can always be traced back to poor/not poor regardless of the ethnographer's pleadings. The firm can use places that demonstrate fast adoption for whatever reason, prioritize some countries over others, and find new ways to encourage adoption where it appears to be lagging. Whereas ethnographic knowledge previously had to be squeezed into a preexisting frame of "emergingness" to create new products that ultimately fit an expected mold, with the TDI it became possible to generate alternative frames.

### The Not-Always-Welcome Overflow of Agency

In this final section, I reflect on the kinds of agency it took to launch the TDI, and specifically the uncertainties at stake in exercising that agency. In Miller's response to Callon (2005), he mentions that market research in multinational firms he studied works only to legitimate decisions already made. This is not an unfamiliar phenomenon, and yet the need for legitimation at all suggests that there are uncertainties in play that researchers might appropriate to create space for more openness. The TDI map in part worked as an "immutable mobile" (Latour 1987), in that it circulates as a finished artifact, exerting its agency without narration from researchers. Yet it managed to do so only because it says what everyone knows already—that something is amiss if Shanghai is in an emerging market—but cannot act on. In this sense, the TDI built on what was there already, but did not merely reproduce the dominant view.

Yet the effort was not just an object that circulated on its own accord. In addition to being a knowledge that walked a fine line between being located in, and dislocated from, "the market" (and anthropology), it was also entangled and disentangled in social connections as it was mobilized by us and others. The data were reworked and contextualized by researchers, new analyses were conducted in response to requests from various parts of the business, but we were wrong to believe that one can control one's own influence through designing objects for others to use. An incident working with a strategist to get the TDI into an executive speech revealed how mutable the index proved to be. For it to be included in the speech—a speech explicitly about showing Intel's customers that emerging markets were worth investing in—a story about what the TDI "really" means had to be created. TDI scores for PCs were growing in the largest emerging markets and were flat in the mature ones. The TDI might be really telling us, for example, that people are spending more of their wealth on PCs in emerging markets. This would be a reasonable conclusion from the data, but I objected on the grounds that it is not news to claim that poor people have to spend more of their income on things that have fixed high prices. Trying again, we came up with new messages, which focused on the increasing recognition of the value of PCs. I pushed to have the executive claim that there were wider, more fundamental changes in high-scoring markets that made the PC possible to adopt. I was overruled. "Recognizing the value" remained.

What happened here? First, the message was not ultimately *my* message. This executive was differently caught up in the politics of expectations. An executive speech in public is a highly ritualized event with highly ritualized ways of talking. This was, after all, a sales pitch for PCs, given by someone whose work it was to convince the industry that the company has what it takes to *act* in those markets. It was not a treatise on market dynamics given by a researcher whose job it is to do high-risk things like producing new knowledge. My speech acts do not have to resonate as cleanly as his, and I can afford misconstruals he cannot. Defying expectations for him risked communicating too much unpredictability, even if the knowledge on which that prediction is based is otherwise sound. The presentation of numbers in speeches also has an affective structure that is directly related

# Design for X

to what numbers people expect to see and which ones they actually see. A business audience will see year-on-year percent growth figures regularly; ratios of ratios (the math on which the TDI is based) they do not yet know how to get excited about. Having to explain what a ratio of ratios is detracts from the affective impact although the news is largely good for profits. Yet when I make this explanation, it *is* the excitement. The method is the new knowledge excitingly not caught up in business as usual.

The certainties of a cleverly designed boundary object were fading. At least in this proposed speech text, western Europe was no longer labeled "mature" but instead "established." It was not a straightforward story of rising income levels in emerging markets spelling good news for the PC industry, but the clustering of countries still followed old mature/emerging distinctions. This was not what I set out to achieve. Was this worth fighting? Perhaps the TDI did not go far enough as a boundary object, yet at this point I lacked the stamina to counter with something more clever, if I could even figure out what that would be. I was beginning to make graphs on the fly, and anthropology provides no preparation for the visual and mathematical craftsmanship necessary to produce them at will. The whole section of the speech was ultimately cut, and I never learned the rationale for the decision.

Throughout the process, I was keenly aware that regardless of what was actually said, I would have been rewarded for having gotten "my" index into an executive speech. Just as I had mistaken the institutional inability to recognize consumers as agential beings for my own lack of agency, it was all too easy to mistake the executive use of this map as confirmation that the world was somehow different now through my efforts. Executive use of this data would be codified in annual performance reviews as a personal accomplishment. However, this misrecognizes the way that agency always overflows the frames in which it is supposed to act. Once the TDI left the mouths of other senior figures, it began to make the rounds in such a way that when it came back, people asked us if we had heard of this TDI thing. This was a sure sign that it had left its "social" origins and was encountered as a disembedded "fact," not a point of view with a social origin. Executives, in this sense, can be thought of as technologies of disembedding. They had the power to make the TDI no longer ours but anyone's. Given the at best multivalent signals being sent in connection with the TDI, what work "I" was really doing through my object was not clear.

Objects do have lives of their own, and their designers do not get to continue to act as their authors. Work in this manner does help escape the overdetermined "design for X" problem, in that X is no longer a homogeneous group with predictable characteristics. Yet precisely because materials are never determinant, no solid grounds exist for claiming what the effects really are. The form it takes—a data set rather than an anthropological narrative—is an open invitation for people to appropriate it and use it to make decisions. I set out to defy predictability by keeping X as X while still enabling choices to be made between different markets to be entered. I invited others to appropriate these new kinds of scales for their own purposes. The TDI informed market selection processes but did not choose them on its own. In no small irony, institutional embeddedness—that Intel already had people in certain countries and not others—often counted for just as much, if not more, in market selections.

In this attempt to design my own work for appropriation, I have learned quite a bit about why design for appropriation is so resisted in user-centered design. It is not merely that companies stand to lose money if they design their products for reuse rather than planned obsolescence. It is also something far more affective, bound up in how one knows that one has had an effect on the world. There are affective satisfactions in drawing lines between producer and consumer and being able to say that *my* product will be valued by half the globe. This is what I suspect the executive I was working with ultimately wanted to say, given the language we settled on. When agencies overflow beyond an object's originators, questions of how agency actually works become matters of concern. In this, interlocutors like me, alongside designers, product planners, and DIY enthusiasts, all have something in common. We may experience our agency as an ability to appropriate what is at hand but maintain the fiction that our work is an immutable mobile while the work of others is entirely manipulable and appropriable. Discursive structures like "design" and "go-to-market engines" feed this fiction by masking the overabundance of agencies at stake.

The overflow of agency can affect the researcher as a person with a professional identity and epistemological commitments. There is a sense in which I do not know which elements of the project I actually believe and find interesting, and which are a necessary performance for an audience. Just as it both embedded me in the business and disembedded me by making my work into common knowledge that circulates without ties back to me, so too it unsettled my relationship with my discipline and points of view as a scholar. Anthropologists are trained to be wary of quantitative knowledge, yet there I am, creating spreadsheets and wondering alongside the product planners why countries ranked the way they did. I picked one anthropology I cared most about—the critique of development—but never sensitized anyone to a particular context. I may have encouraged conditions more conducive to it, but this is not the same thing and is in any case only a post hoc remembrance. Only by dealing with these numbers in a hands-on way do I now see them as something more than a concession to what the business will value. Just as the TDI created more possibilities for other people to become attached to the project, it is not possible for me to think of it as merely a boundary object, "only" an invitation for others to begin engaging in the discursive critiques of development that I "really" care about. In my roundabout attempts to pry open a system of production and destabilize it, I have unwittingly enrolled myself in its hybrid, shifting epistemics, multiplying the gaps and disagreements not just for others but for myself as well.

Design so often implies something cynical, as in the phrase "to have designs on." As Ingold (2010) argues, design implies a clarity of intentions that do not move as a result of the act of designing and carrying out the design. This could not be further from the case with my work on the TDI. An academic debate about what is and is not embedded, and how things come to be abstracted from social relations, involves more than just theories or an enactment of those theories in the "real world." Questions of what is or is not embedded become profoundly vexed ways of being. This vexing comes all at once, contemporaneous in one single act. Letting go of this or that entanglement is not as cost free a process. I have long gotten over tedious hand-wringing over whether working outside an academic department constitutes selling out or a betrayal of supposedly pure critiques of capitalism. Long ago Donna Haraway (1991) killed off any pretense that laundering one's salary through this or that institution makes purity an achievable, or even desirable, state. Yet what is left is still a problematic sense of agency and the disquiet that comes through generating new social configurations that may ultimately turn around to exclude the very sensibilities a researcher might care about.

Many other contributors to this volume have designed objects to work in similar ways, interjecting unpredictability through their creations, which themselves can be subverted. In my own campaign against rendering consumers predictable, it became all too clear that openness is a two-way affair. To work in this way requires both researchers and "real-world" actors to open themselves up to the indeterminacy of agency and its haunting multivalence. There are research payoffs in taking these risks, of course; designing things that you can then follow is a reasonably good research strategy. Yet to explain it away as merely an investigation would be a lie.

Callon treats anthropology as a general term to mean a social theory, not necessarily the theories that come from the discipline per se. He asks, "How can one ensure that the success of Linux and its anthropology of applied [economics], which is causing Microsoft so much concern and forcing it to alter its strategy, does not remain a miraculous exception served by exceptional circumstances? Answer: by facilitating access for all anthropological programs. ... Establish a right to experimentation and to discussion of the results obtained" (2005, 13). The equivalence that Callon makes between the social theories posed by the open-source world, those proposed by academic anthropology, and those proposed by neoclassical economics are not, in my view, equal substitutes competing for successful adoption. The playing field is in no way level. That they might be competitors on an open market suggests, as Miller charges, that Callon has embraced the neoclassical laws of the market a little too much. But the wider point holds. All kinds of anthropologies can shape markets, as well as be shaped by them, through experiments that are suffused with unpredictability. Experiments, through their very unpredictability, do open up the possibility for a different social imagination to take hold. Perhaps—and only just perhaps—whoever X is, and what he or she does, might not have to be so prescribed.

## Acknowledgments

While this account is my own, ken anderson and Phil Howard were equal contributors to the TDI project. Maria Bezaitis, Tony Salvador, Renee Kuriyan, Jim Galanis, and Herman D'Hooge gave vital insights without which we could not have proceeded. I would like to thank James Leach, Lee Wilson, Daniel Jaffee, Tad Hirsch, Suzanne Thomas, and Daniel Miller for comments on earlier drafts of the chapter. All errors of fact or interpretation are my own.

## References

Burrell, J. 2011. User agency in the middle range: Rumors and the reinvention of the Internet in Accra, Ghana. *Science, Technology, and Human Values* 36 (2): 139–159.

Callon, M. 1998. *The Laws of the Markets*. Oxford: Blackwell.

Callon, M. 2005. Why virtualism paves the way to political impotence: Callon replies to Miller. *Economic Sociology: European Electronic Newsletter* 6 (2): 3–20.

Dourish, P. 2003. The appropriation of interactive technologies: Some lessons from placeless documents. *Computer Supported Cooperative Work* 12 (4): 465–490.

Ferguson, J. 1994. *The Anti-politics Machine: "Development," Depoliticization, and Bureaucratic Power in Lesotho*. Minneapolis: University of Minnesota Press.

Haraway, D. 1991. *Simians, Cyborgs, and Women: The Reinvention of Nature.* New York: Routledge.

Horst, H., and D. Miller. 2006. *The Cell Phone: An Anthropology of Communication.* London: Berg.

Howard, P., K. Anderson, L. Bush and D. Nafus. 2009. Sizing up information societies: Towards a better metric for cultures of ICT adoption. *Information Society* 25 (2): 208–219.

Ingold, T. 2010. Designing environments for life. Seminar paper. University of Aberdeen.

Latour, B. 1987. *Science in Action.* Cambridge, MA: Harvard University Press.

MacKenzie, D., and J. Wajcman. 1999. *The Social Shaping of Technology.* London: Open University.

McDonough, W., and M. Braungart. 2002. *Cradle to Cradle: Remaking the Way We Make Things.* New York: North Point Press.

Miller, D. 1997. *Material Culture and Mass Consumption.* Oxford: Blackwell.

Miller, D. 2005. Reply to Callon. *Economic Sociology: European Electronic Newsletter* 6 (3): 3–13.

Nafus, D., and K. Anderson. 2006. The real problem: Rhetorics of knowing in corporate research. In *EPIC Proceedings.*

Plattner, S. 1989. Introduction: Economic anthropology. In *Economic Anthropology*, ed. S. Plattner. Palo Alto: Stanford University Press.

Schor, J. 2010. *Plenitude.* New York: Penguin.

Slater, D. 2002. From calculation to alienation: Disentangling economic abstractions. *Economy and Society* 31(2): 234–249.

Starr, S. L., and J. Griesemer. 1989. Institutional ecology, "translations," and boundary objects: Amateurs and professionals in Berkeley's Museum of Vertebrate Zoology, 1907–39. *Social Studies of Science* 19 (3): 387–420.

Strathern, M. 2002a. Abstraction and decontextualisation: An anthropological comment. In *Virtual Society? Technology, Cyberbole, Reality*, ed. S. Woolgar. Oxford: Oxford University Press.

Strathern, M. 2002b. Externalities in comparative guise. *Economy and Society* 31 (2): 250–267.

Strathern, M. 2004. *Partial Connections.* Oxford: Rowman & Littlefield.

# 11  Engaging Interests

Marilyn Strathern

This volume takes its own inspiration from the extent to which inspiration comes from people's ingenuity in making new tools out of old.[1] If I am puzzled by why that should be so compelling a theme, it is a puzzle that turns out to be quite a good place to begin.

## A Question

Why should the thought of adapting techniques, making new from old, cherishing the potential for change by making allowances for it in advance—why should these aspirations indeed seem so compelling? More than just compelling, why should they touch or amaze or enchant us, in short, carry affect? There is something heartwarming in the engagements so invited, so freely opened up.

It can't just be the examples to be found in these chapters—though I observe they tend to keep away from military devilment, instruments of torture, and the like. It can't just be the value we—we whom the editors have called together here—put on ingenuity or creativity or novelty. For we—who the editors make clear belong to an epistemic community that debates and criticizes these things—have learned to be wary of the productionist paradigm that sees a virtue in everything pressed into service, or uses usefulness as a universal measure, as we might be wary of any simple formula to anticipate potential by design. It can't just be the aesthetics of complexity, and the delight in knowing that one can will (that is, intend) or create the conditions that will guarantee that we never know everything anyway. So where does the compelling nature of many of the examples discussed here come from?

Perhaps we can find a clue in the way the original conference was convened. One of the generative features that the conference program had built in was that there simply could not be any single position that would characterize all the interventions, no summarizing value by which to measure the intellectual capacity and capability the conveners hoped it would yield. My inclusion was an example. In extending their invitation, they were including someone whose interests touched on several of theirs at certain points, but who could in no sense presume to have grasped the scope and scale and detail of the ideas concerning ICT discussed here. The compulsion for me was to align my own interests in such a way as to create a background commentary that might have some purchase. Let me take that compulsion as a truth about the engagements that are happening here, in this volume, and one worth pondering.

I wonder whether the contributors are not, in fact, dealing with some ancient social capacities that are themselves made anew in each act of communication or technical deployment. Is the affect also a frisson of recognition? Is this in part what makes the topic of the volume so compelling?

**Interests**

Spelled out, the social capacity to which I refer sounds rather banal. But perhaps what is recognized afresh each time entails an appreciation of potential, of anticipation, namely, that engagement as such is possible. And engagement in turn is possible because *people see that they—themselves, others—have an interest* in the collaboration or exchange or exploitation or whatever.

As to the notion of interest, many of the research interests represented in this volume have run parallel to, and veered away from, the popularity of intellectual property over the last two decades to simultaneously establish conditions of innovation and exchange of ideas side by side with conditions of property ownership. And if property is about anything, it is about interests. But whose interests? The question has been asked countless times, not least by the Royal Society (Britain's academy of science), whose review in 2003 pleaded that we need to encourage an appropriate environment for fair exploitation, not focus on who owns the IPR.

But while IPR can be left behind, it would seem that the issue of interests cannot. Some of the chapters here directly ask whose interests are at stake. So this is an issue that endures beyond the forms of property claims.

If I implied that in extending the invitation to include me, the editors made me a momentary extension of their concerns, I did not mean automatically so. When persons are added to people's enterprises, they have already been moving in diverse spheres of their own, and they come with various intentions and motivations already embedded in what they are thinking or doing. Indeed, that is often their value, as the addicts of networks will tell you. One may even want to *foster* the distinctiveness of others' interests, as happens when people are encouraged "to take an interest in" their material heritage, or instead downplay them in avoiding "conflicts of interest."

This is where one would give a social response to the editors' question of whether diversity matters. It is in everybody's interest to move in a world of multiple interests. There is no problem, it would seem, in conserving that diversity.

One mechanism lies precisely in making people *extensions* of other people. The social contexts in which this happens can be created regardless of any declaration of common concerns, regardless of how large or small the numbers, how selfish or altruistic (if I may use the shorthand) the motives. What I mean is quite simply that persons see that what other persons do holds an interest for themselves. And they bring it toward or reach out beyond themselves by engaging these others in their enterprises. Again, banal. We do it without thinking. What *might be* given thought is just how relations can be designed, can be kept open enough, so that there is a potential for the locking of interests even though they cannot be specified in advance.

In fact there are places in the world where people go to great pains to guarantee that interests will be divided. An example familiar to social anthropologists comes from Hagen in the highlands of Papua New Guinea.[2]

## Divisions of Interest

Think of the rule of exogamy, the stipulation that you should marry outside a particular group, such as a clan. Exogamy simultaneously defines sets of people as insiders (those who cannot marry one another) and as outsiders (those from whom spouses come). Taking up the vantage point of one's clan group makes this clear. But note the way perspective folds back on itself. If another clan is an outsider from *the point of view of marrying* them, then they are also brought within. A woman leaves her own clan. In local Hagen idiom, she was referred to as being "severed" from her clan of origin to "go inside" the husband's. Meanwhile her brother takes a wife from an outside clan and, through her outside powers, produces further clan members. If pressed to the point, highlands people would probably say that it is only by bringing in external sources of fertility that a clan can reproduce at all. And that external source is kept external. Although a woman "goes inside" her husband's clan, she does not lose her connections with her own kin;

far from it, she becomes a road between them, so that her external origins are conserved. Rather than fully absorbing her, the new clan body encloses or encases her. Men routinely refer to women as "fenced," as though one could imagine a small enclosure within the clan territory surrounded by the wider perimeter of the clan land, an image not unlike that of gardens that are individually fenced (against predation from domestic pigs) within a clan territory marked by boundary plants.

The flow of bodily substance between persons is also a process of extraction or elicitation. What the woman takes within her, she later brings forth as children. Those transfers of bodily substance have a counterpart in various artifacts that also flow between people: there is a whole mimetic that runs alongside bodily flows made visible in the transfer of material items. Conversely, the substance flows give life to the exchange of artifacts. Now, this counterpart flow of artifacts is both available for independent manipulation and locked into body process in a direct way. The items I am referring to have value because they are equated with different aspects of the body and are regarded as being extracted from persons in just the same way as persons (like the bride) are regarded as extracted from other persons. In fact, the two systems interpenetrate precisely at the point of marriage. In arrangements of the kind I am talking about,[3] it is these external artifacts that *elicit* persons. One clan yields its daughters (in marriage) to another because that other clan has offered artifacts (bride wealth) in the form of valuables for her. Valuables these days include money and a living symbol of wealth in the form of the principal domestic animal, the pig.

The rule of exogamy, then, means that husband and wife facing each other across a hearth do not just come from different social groups; they are explicitly divided from each other by their interests, that is, by their loyalties and orientations. I am not going to call exogamy a "technology." I am going to point out that it is set of protocols that work, that is, a set of rules for action. They have as their purpose the engagement of people's interests, producing marriageable and reproductive pairs and reproducing those interests themselves as always in divergence.

The point was made earlier that thought might be given to just how relations can be designed, can be kept open enough, so that there is a potential for the interlocking of interests although they cannot be specified precisely in advance. It turns out that there is an example in this rule of exogamy, and it is not confined to the highlands. What might initially strike anthropologists as a rather fixed practice in terms of culture and tradition informs a social practice that holds enormous possibilities for innovation. The innovation I have in mind is one that has come about with Papua New Guinea's internal globalization; while it could not have been anticipated by these protocols, that process is nevertheless much facilitated by them. These days people marry across cultural and "ethnic" divides all the time. Local migration, urbanization, and endless traveling all contribute to what they call "mixing." Yet the *expectation* of division between intermarrying sets of kin, to the point of negotiating whose bride wealth practices are going to operate, renders it all surprisingly unproblematic. No worries about ethnic mixing, at least on this axis!

Protocols put to new work. I see an interesting point of contact here with many contributions to this volume, and with what is called, in a most general way, technology.

## Technology

In his *Technologies of Choice*, ten years before the Royal Society's pronouncement, Pierre Lemonnier (1993, 22) says that technical invention remains one of the great puzzles. The issue is how people perceive a sufficient gap in what they have to hand to want to plug it. If a system works, how can that conceptual gap arise? How can one conceive of something one does not have? His particular interest then becomes why some technical practices are chosen over others. Why should people wish to improve on what works? In fact, part (only part) of the answer is already contained in Bruno Latour's contribution to Lemonnier's volume: from some points of view, this is a false model of the relationship between artifact and use, even as questions about cultural elaboration are a false model of the relationship of (useful) artifact to (ornamental) object (378).

Rather charmingly, in his paper on Aramis, the failed rapid transit system that was to have produced a new underground network in southern Paris, Latour also introduces an ethnographic pig, specifically from Papua New Guinea. He is warning against the epistemological dangers of seeing the pig as first an animal and then an item of cultural value. That takes us to inscription as the archetypal "cultural" activity. Latour develops his countermodel of quasi objects as technique and sociality enfolded inseparably together.[4] The anthropologist should be no less adept than his or her informants at taking on other perspectives. Starting with the highlands view, then, one might well wish to ask about quasi persons: persons have relationships enfolded within their bodies,[5] at once exterior and interior to themselves. And the artifacts they make (including the pigs they rear, and in the past the stone axes or shells they exchanged) have similar qualities.

Pigs remind us that people have bodies. The interests that people conserve and reproduce through their social relationships provide them, in their dealings with one another, the bases for countless cognitive operations—representation, anticipation, reflexivity—that work through the way they conceive of the disposition of bodies in relation to one another.

As a living organism, the body constantly, iteratively, processes materials, ingesting and egesting. Such iteration is not mere repetition. The body is in a constant state of fresh animation through what it takes in and gives out. Perhaps those rhythms are a component of the kind of anticipations we find in social and cultural practice: the ability to repeat actions in other registers is a reflection (and enactment of) the replenishment of energy.[6] The point is that renewal of energy or capacity is not confined to the repetition of the same tasks (replication) but can be equally well renewed in the ability to transform one task, or relationship, into another (reproduction).

Now, the question of why people should wish to improve on what works contains within it a reference to technology with which I want to end. As has already been remarked in this book, *technology works*. That is what it is, or it is not technology. There is a kind of intractability that characterizes the making of "things."

The challenge, it seems to me, in many of the chapters here is just exactly the manner in which technologies, things that work, in lending themselves for new work, come to work in different ways. The editors refer to rewriting or reconfiguring, and what matters is not identity or classification, whether this lump of metal still is or is not the original water pump or whatever, but whether the lump of metal or software filigree has *work to do* (for whatever reconfigured purpose). Against all the imagining of its uses lies the intractability of the technological demand that the innovation somehow "works"—otherwise it is not a technological innovation.

It might be helpful in contemplating the many applications of communications technologies to think of the notion of "interest" as a kind of social counterpart to "technology." As people put energy into what they make or what they do with what they find, they also come up against the intractability of the interests that compel the actions in the first place. As Latour (1993, 391) observed: "An object cannot come into existence if the range of interests gathered around the project do not intersect." Unless interests are enrolled, nothing will happen, for if interests are not satisfied or not made evident, then they have not been made present. An enterprise must capture interest; otherwise it serves no purpose and is not an enterprise.

## Notes

1. This afterword is kept here in a form close to the original address to the workshop "Subversion, Conversion, Development: Public Interests in Information Communication Technologies," convened by Robin Boast, James Leach, and Lee Wilson at the Centre for Research in the Arts, Social Sciences, and Humanities, University of Cambridge, in late 2008.

2. I deliberately add to elements in the editors' introduction here.

3. Arrangements of this kind are not universal in Papua New Guinea by any means. Some of the controversy is adumbrated in Godelier and Strathern 1991.

4. He has a wonderful passage about the insouciance with which Euro-Americans identify "more [a higher level of] society" with "more technology" (Latour 1993, 380). (The definition of human society becomes the more certain the more they can identify the enlistment of nonhumans—tools, artifacts, plants—in people's interactions.)

5. Although I have not brought this into my exposition, throughout Melanesia one cannot talk of bodies without also referring to mind, intentionality, vitality, and animation.

6. With the Melanesian material, we are in a world where people do not just reflect on their activities but are reflexive about them, able to switch perspectives, anticipate outcomes, and see the past not as tradition contained in the present but as an eversion or transposition of the present.

## References

Godelier, M., and M. Strathern, eds. 1991. *Big Men and Great Men*. Cambridge: Cambridge University Press.

Latour, B. 1993. Ethnography of a "high-tech" case: About Aramis. In *Technological Choices: Transformations in Material Cultures since the Neolithic*, ed. P. Lemonnier. London: Routledge.

Lemonnier, P., ed. 1993. *Technological Choices: Transformations in Material Cultures since the Neolithic*. London: Routledge.

Royal Society. 2003. *Keeping Science Open: The Effects of Intellectual Property Policy on the Conduct of Science*. London: Royal Society.

# 12 Subversion, Conversion, Development: Imaginaries, Knowledge Forms, and the Uses of ICTs

James Leach and Lee Wilson

Infrastructure ... never stands apart from the people who design, maintain and use it.
—Geof Bowker, "The Knowledge Economy and Science and Technology Policy"

Lievrouw and Livingstone have defined (or redefined) new media as "information and communication technologies *and their social contexts*" (see Lievrouw 2011, 7; our italics). In doing so, they emphasize that new media are interesting and important inasmuch as they combine three main elements: artifacts or devices, practices, and the arrangements and social forms built around practices. "Today, a lively and contentious cycle of capture, cooptation and subversion of information, content, personal interaction, and system architecture characterizes the relationship between the institutionalized, mainstream center and the increasingly interactive, participatory and expanding edges of media culture" (Lievrouw 2011, 2).

In this, a volume that combines a focus on apparently attractive phenomena and their combination (ICTs and local/indigenous knowledge; interactive, participatory, and emancipatory social movements), it has been important to maintain a critical stance, and a reflexive one. Turnbull and Chambers write of the various "seductions" of both indigenous knowledge systems and ICTs and usefully caution us against leaving our critical faculties behind at the door, as it were, in both cases. Strathern too highlights this concern when she asks questions of the overall theme. Without prejudging an answer, she wants us to consider why we should be so interested in the adoption, repurposing, and unexpected development of technology as reflecting local politics, culture, or development aims.

For one thing, rather surprisingly, while global transformations wrought by ICTs are widely trumpeted, we have remarkably few accounts of ICT initiatives that have meaningfully privileged local knowledge and understanding. The reciprocal effects on the development of these technologies and accompanying social forms in different contexts remain underrepresented.

While authors in science and technology studies have examined the relationship between subjectivity and materiality in considering the ways in which power and authority are embodied in technological objects (Winner 1980, 1989), accounts that consider producer-consumer relationships, how these might transform over time, and, critically, be transformed by the analytic approaches used to engage and understand them, are limited in number. Notable in this respect, though, are Neff and Stark (2003), who demonstrate the ways in which values encoded in information technologies work to affect organizational and economic processes, and Woolgar (2002), who draws attention to the significance of the "real" in relation to "virtuality" and the overemphasis of the transcendent nature of ICTs in much analysis. Following these insights, our contributors have focused on the real and actual potential in specific situations of ICT-fueled social developments.

Given that many uses of ICTs allow the visualization and representation of knowledge, it is unsurprising that several contributors focused on mapping and its socially transformative possibilities. Nafus, for example, charts her development of a map for Intel executives showing technology uptake across the globe, a map that pretended to universality. Her brilliant intervention, producing a map that destabilized the familiar narratives and perceptions of planners and designers, consciously took the form of a "panoptical rendering" of the world for the specific reason that to have an effect in *that* context (and a potentially subversive or developmental effect), information had to be rendered in a manner that spoke to the interests and values of the senior executives she was trying to influence. "This disembedded 'god trick' (Haraway 1991) is precisely what situates it within an industry that explicitly eschews designing technologies for a particular place," as she explains. In doing so, Nafus points to two important wider conclusions from these chapters: first, there is a universalizing and flattening tendency in the production of ICTs; and second, mapping and forms of visualization can work as a mode of empowerment against the grain of this effect. At one and the same time, she illustrates that mapping can be transformative by remaining fiercely local and relevant locally (Lewis).

The volume as a whole highlights how ICTs may be engaged both to elide and to facilitate different interests (Strathern). As Verran and Christie put it, "As long as we do not make assumptions based on modern ways of using digital objects, if we proceed in open ways, collaboratively and empirically researching how indigenous people actually use digitizing technologies, we keep open the possibility of strengthening traditional forms of cultural innovation with computers." Both Bala's and Diemberger and Hugh-Jones's chapters (as well as those of Petersen, Lewis, and others) demonstrate the truth of this possibility admirably. Diemberger and Hugh-Jones explain that knowing what a book is, or what a digital rendition of a book may be, is not predictable across contexts and localities. Verran and Christie show that for Aboriginal people, the status of digital objects such as databases is "difficult to tell." Not only may the inhabitants of Orkney described by Watts have little use for ubiquitous communication, but to provide it may also trample on an important local sense of uniqueness, one that impels innovative and collaborative adoption of other more appropriate technologies. Where users are "surprising and unpredictable," it is incumbent on us, the authors here argue, to tread carefully. As Verran and Christie write, developers and implementers of technological (and development) solutions need to be "aware of how easily other lifeways are crushed by modern ways of going on." We are shown clearly that imaginaries that shape technologies and their uptake are always placed and based on assumptions emerging from that placement.

There is then a theme, articulated and represented here by many and already alluded to in the opening chapter, that all knowledge is local knowledge. Indeed, at the opening and the conclusion of the collection (Watts on the technology industry's relation to a rural hinterland, Nafus on the business of selling technology around the world), analysis reveals the locality of imagination, of interest, and the importance of an impetus to replace the all-too-common perception of a placeless and universal desire for communication, to rethink ICTs as somehow neutral in facilitating the desire for communication. The contributors emphasize how the cultural and social assumptions that people hold define the production of technology and how it is used and modified by people who are often differently situated.

Verran and Christie give us the term "imaginaries" and point to their importance in structuring relations. "Working imaginaries are narrative and image, metaphor and analogy. They frame and explain; they are stories and pictures that figure, prefigure, and refigure relations." They emphasize the importance of the imaginary as the structure in which information and knowledge find their (particular) purchase and relevance. Watts describes the common imaginary of the technology producer, an imagining of a kind of universal future colonized and standardized by the communication devices they produce. She links this to a particular landscape, and practice within it, and contrasts that to a different landscape in which this imaginary makes less sense. Verran and Christie's chapter emphasizes that the database *is* a local knowledge practice for exactly the same reason (see also Strathern 1995).[1] It embodies a metaphysic and worldview, one closely allied, we might say, to the modernist impulses of the technology planners that Nafus writes about. All these positions demonstrate particular interests, in the case of technology producers, often elided as recognizing and adapting to universal human needs or desires.

While the standard ICT industry model has been to anticipate the social as a market to be addressed and manipulated, building obsolescence into the objects it manufactures to stimulate desire, the contributions here have highlighted a different set of possibilities driven by a different political economy of sustainability. The simplicity of the idea of reuse from the free and open-source software movement has been inspirational for many, and in different ways, around these contributions.[2] And in truth, our motivation to attend to the possible reuse and conversion of information and communications technologies in this volume has its foundation in the "enablement" (Corsín Jiménez 2011) potential of technologies that embody both material and political/ethical futures (Leach 2009; Lievrouw 2011, 24). With the right conditions, technologies can be "wild things," as Pedersen describes the particular assemblages that stimulate the human tendency to recombine and repurpose both things and ideas. Such conversions, gathering interests (Strathern), at times subversive, are what Lievrouw terms "*reconfiguration*, where users modify and adapt media systems as needed to suit their various purposes or interests" (2011, 4).

Engaging with an important metatheme also central to our introduction, Mark Deuze (2006) cites bricolage as one of the main (three) elements in engagements with new media. Focusing on what Lievrouw (2011, 4) terms "remediation" (i.e., content and "meaning making"), bricolage is an important process because it "describes a constructive and interactive participation" (Deuze 2006, 66). The chapters you have read are, of course, about meaning making. (What human activity is not?) However, they make a distinctive contribution in line with our argument in the introduction about the importance of the particular improvisational parameters of the social forms in which they are situated. Such improvisations amount to an active and ongoing shaping of the technologies and infrastructures *themselves*. While new media theory focuses on the possibilities for subversion and social development by mix and remix at the level of meaning and content, our intention in presenting the cases in this volume (starting with Rai Coast dogs' teeth and Levittown) was to prepare the reader for a shift in focus that the contributors make: a shift to confronting objects as if they were open and reconfigurable. That is (as is abundantly clear) because of an interdetermination of the technological and the social in their examples (the combination of material, practice, and institution that Lievrouw and Livingstone highlight in their notion of "recombination").

Having come this far, readers will now know that the contributors have developed and elucidated different approaches to, and understandings of, the place and potential of ICTs building on exactly this realization.[3] Technological innovations do not just act on the social world to effect change; they are themselves a product of particular social relations. They embody, reify, and articulate these relations. Their "enablement" function often has a transformative power. Through sustained ethnographic analysis, the contributors have worked to decenter the design process by variously revealing, analyzing, and documenting how these particular reifications come to shape use, conversion, development, and transformation. In doing so, they open up a space for critical reflection on processes of social and technological change. This is a space in which we emphasize use and user development. They offer valuable insight into community-driven initiatives working to realize the potential of new technologies to redress global imbalances in the digital knowledge economy (cf. Atkins, Brown, and Hammond 2007) because they have deliberately located their accounts of "technological" objects within a sense of the possibilities of reconfiguration. We think that they offer a unique consideration of the inflections that cultural expectations and values place on the things people use and how they come to adapt them. In this sense, the chapters present a timely insight into participation in the digital economy and the mediation of experiences of modernity through the adaptation and appropriation of (sometimes new) technologies to suit the exigencies of real use. It is worth reviewing them in a little more depth to draw these aspects out explicitly.

In chapter 2, Watts introduced us to a way of engaging the complexities of the encounter of imaginaries through a different way of presenting what people know and how we might represent that in a way true to the whole. She does this explicitly because so much can be lost in the conventional descriptions-made-real by configurations of politics and the interests or assumptions of developers. In chapter 3, Petersen involved the reader in a particular configuration of the use of ICTs in a political project of social transformation. The possibility and hope generated by these media are significant elements in how this development of community took place. A version of enablement was outlined that draws directly on the sociotechnical possibilities of particular objects. Having tools with such potentials adaptable and accessible as modes of engagement gives ethnographic substance to the idea of possibilities, but it also requires a consideration of the different epistemological commitments of users and adapters. Verran and Christie then explored this idea by turning to digital archives and databases. In this and the next chapter, we were challenged to think out the structuring of knowledge when long-standing practices of ecology, or spiritual practice, are adapted to digital preservation or presentation. It was clear how presentation either succumbed to, or needed to modify, modes of organizing and making information available as a material aspect of practice, not just of its representation.

As Bowker and his coauthors write:

> Research on distributed knowledge processes suggest[s] a critical conflict between knowledge processes in groups and the technologies built to support them. This conflict centers on observations that authentic and efficient knowledge creation and sharing [are] deeply embedded in an interpersonal face to face context, but that technologies to support distributed knowledge processes rely on the assumption that knowledge can be made mobile outside these specific contexts. (Bowker et al. 2000, 1)

Demonstrating the importance of understanding particular knowledges as different modes of organization and engagement to make them comprehensible followed in the chapters of Diemberger and Hugh-Jones, Lewis, and Chambers and Turnbull. They offer a consideration of some of the practical, intellectual, and philosophical resources that might help us understand the meeting of different interests and values. Their case studies ask us to consider how we are to give the space for the intricacies of knowledge forms to exist as coherent and effective entities when realized with ICTs. The stakes for development, as well as the political-cum-ontological stakes, were made clear.

As a case in point, we were shown that, as with the Baka and Yaka peoples that Lewis describes, the assumption (for example) that visibility, distribution, and access are always positive, and communication is always desirable, is a partial and interested account of human possibilities and futures. Information does not always want to be free because information is an inextricable part of knowledge. And knowledge depends on and refers to the specific relations between particular people. People do not always desire more communication or more connectivity, although, as evidenced in the chapters, they usually do want more autonomy, recognition, and control over the conditions of their existence. That can be achieved by allowing more space for repurposing, for technologies built to be more adaptable, convertible, and open. That requires reflexivity as well as understanding.

From the accounts presented, we understand that knowledge, and hence knowing, is not singular, nor is it determined from an authoritative center. It is multiple, local, and diverse. Knowledge in the conception of our contributors can be thought of as those practices, certainties, stories, and imaginaries that are held and maintained by groups of people. Given these contributions, we hold that ICT development—in both education and civil society—should adopt and enable diversity of use, local modification, creativity, and significance. Such an approach is long overdue. But what does it mean in practice? The chapters here give us concrete case study examples and analysis of those examples to answer this question. We tease out some of the answers in the following sections.

## Development and Users

With the spread of digital media and the importance of structured computer programming (Blackwell), questions of plurality and convention in design, production, and use have become especially salient. Developers recognize the need, in fact, to strike a balance between standardization and flexibility (Hanseth, Monteiro, and Hatling 1996). Such tensions, while not in any way new, are especially pronounced in increasingly complex technologies that demand compatibility if they are to achieve broad use (or indeed market domination). Subversion of purpose, principle, and design to some extent entails a movement against a prior state of affairs. In this respect, inasmuch as designing for purpose is a normative practice, the practicalities of cultural improvisation in their specific manifestations can work to (re)define the politics of design.

Our use of the term "development" is intended to highlight how the chapters demonstrate a space around the encounters between knowledge systems and their adjuncts in technologies, a space in which development in the narrow sense (of changes that have a rationale or reason) are occurring. There is, quite simply, a development of technology in several cases here. In others, there is a space in which social, political, and technological developments could and perhaps should occur. Further, we have been given examples of how ICTs can, when converted, collaboratively developed, and iteratively modified, make development processes (this time in the sense of "Third World development") more appropriate to each circumstance and mitigate the destructive influence of rapacious appropriation or market domination. Both Lewis and Bala, for example, demonstrate that the social/technical adoption of appropriate technologies is good for both local interests and business. Far from using such technologies to resist development, people are adopting and modifying ICTs to enhance development opportunities in novel and unexpected ways when conversion occurs.

The potential for new technologies to bring about social good is obvious.[4] Yet the implicit values and assumptions that inform the design of

ICTs are often neglected when they are applied in social and economic development. Indeed, there are compelling similarities between the development and uptake of new technologies and the success and failure of economic development initiatives. Generalizing to make the point: in both the design and implementation of ICTs and development discourse, one often finds the rhetoric of user-centered design, or local context. Yet both product development and development interventions often fail to achieve their stated aims precisely because they are driven by imperatives that are alien to, or at odds with, their perceived value to users. The space for conversion is paramount, one might say, to the possibility of success.

Practitioners of participatory approaches to design are keenly aware of the politics of production and the ways and means through which stakeholder participation might be actively facilitated (e.g., Cahill 2007; Kensing and Blomberg 1998). Professionals working in fields that demand intercultural collaboration have sought to bring together insights from computer-supported cooperative work, science and technology studies, and postcolonial studies to account for dynamic cultural processes and power relations that frame systems design in culturally complex settings (Irani and Dourish 2009). While these approaches emphasize the role of the user as agent of technological change, the contributions here add the capacity to articulate instances of use driven by quite different modes of critical thinking about, and being in, the world. The growing literature around the notion of user-centered design and participatory technological development (e.g., Downey and Dumit 1997; Cefkin 2010) tends to treat user engagement only as part of an initial inquiry into what ought to be built, a segment in the process of design that subsequently happens elsewhere. The chapters here provoke us to consider a more radical engagement, one that subverts the idea of the single kind of user, or for that matter designer, and the desire to predict or meet the needs of the end user through products that all too rapidly become obsolete or overwhelm local values. Recent work in design studies has increasingly stressed such an approach. For example, Redström suggests the consideration of "how," not "who," in design. Thus user-centered design might be reinterpreted as the exploration of discrepancies between "definitions of use through design" and "definitions of use through use," or between those representing the design domain and the domain of intended use (Redström 2008, 414–415). The questions one might ask following this are seen "primarily as a matter of the relations between different acts of defining use and, thus, not necessarily as a matter of designers' and users' 'roles'" (421). We would argue, though, stimulated by the contributions gathered here, that answers to "how" questions are ultimately underdetermined (Quine 1960, 1–73) and have hinged on a consideration of, rather than a collaborative engagement with, users.

As an example, we note that in the field of the use of ICT for development, the importance of local knowledge to the success of interventions is broadly acknowledged (see Krishna and Madon 2003). Yet the design and implementation of new technologies are driven by values and priorities incommensurate with the underlying cultural logics and values of users and fail to adequately consider the practical conditions that pertain to the local use and uptake of ICTs. As several contributors emphasize, a collaborative engagement with users "requires bringing two very different worlds together on relatively equal terms" (Lewis). This condition is addressed directly by Verran and Christie in discussing the practical possibilities of, and problems with, the databasing of Aboriginal knowledge; in Turnbull and Chambers's discussion of the equivalence and value (if not commensurability) of different knowledge systems; and in Nafus's project to bring the agency and diversity of consumers into the planning stages for technological development.

Lewis shows what open technology looks like in practice. The iconic symbols he discusses are adaptable to different social and cultural contexts and reinvigorate the existing technology of GPS mapping with business and conservation relevance. As he notes, these "maps become ... emissaries, able to communicate ... pressing concerns to powerful outsiders in office-based meetings to which Pygmies would never be invited." To achieve this end requires not a consideration of users but an iterative and reflexive engagement with them in the design process. It also requires that we take into account the context and reason for their engagement.

Bala provides an illuminating difference to Lewis's paper (where open development, as a design process, incorporates anthropological understanding). Bala's is a story of the apparent usefulness of ICTs, off the shelf, as it were. But this is not as straightforward as it might look. In her story, ICTs are also used in a particular way in a specific social and political context. They may not be physically reconfigured or developed, but they are intelligently positioned as (converted to) markers in wider political struggles. To adopt ICTs for Kelabit highlanders is to find new routes to traditionally valued social ends. Their potential and use are shaped according to those ends, and the underlying values they demonstrate. Thus there is a close correlation between the story Bala tells and the other chapters. They are primarily about the rich social embeddedness of use and adoption.

## Ways of Knowing and Modes of Life

Returning to the invitation to reflexivity with which we began this conclusion, the chapters are clearly not (only) about the virtues of ICTs as they spread and emancipate; they also partake of a fascinating moment when historians and philosophers of science turn their attention to the practices and status of indigenous knowledge systems. In turn, anthropologists have turned their attention to scientific knowledge practices and knowledge production in capitalist economies. An idea linked closely to this scholarship is the common understanding, articulated here by Verran and Christie, and Turnbull and Chambers (and followed by the anthropological contributors), that the claims of science to universal validity are usefully balanced by ethnographic attention to the actual, situated production of scientific knowledge (e.g., Latour 1987; Bowker and Star 1999; Gallison and Daston 2007; Stengers 2005). This in turn stimulates a reappraisal of the validity of local and indigenous knowledges as an analogous, rather than inferior, kind of knowing. The theme here is again that culture is not separate from technology (Lemonnier 1993), as the authors insist that the modernist paradigm of a rigid and exclusionary divide should not continue to pervade (at least the ideology) of technology production (Latour 1993). What we see in the chapters is an emphasis on knowledge as something that people do, live through and by, and thus as something that is complexly related to the potential of ICTs to capture, store, reveal, and disseminate information. The concern with knowledge and its value is thus also with its possible rendition or distortion (Bowker and Star 1994). Another entry point to this is the useful introduction of metaphysics, the presuppositions or "heuristic principles" that define forms of enquiry (Collingwood [1940] 1998) and structure the way knowledge is presented and recognized, or structure the very technologies through which those representations can be made. The chapters here emphasize this idea with their repeated links to place, locality, and the situatedness of social, political, and technological imaginations (cf. Woolgar 2002).

There is an important distinction between knowledge as a representation of the world and as a performance of the world. The two are bound up in the idea of a metaphysics both of knowledge and of technology. While technoscientific knowledge has an ideology of representation (what it knows is not the world but a representation of it; see Blaser 2010, 18–21), in practice the ways in which its knowledge is ordered, institutionalized, and made available (through technology) make it a performance and intervention as well. Lewis, Bala, and Diemberger and Hugh-Jones all show clearly that for many adopters of technology, it is the ability of ICTs to reinvigorate their performances of their worlds, and to index their agency in them (Gell 1998), that motivates not just adoption but, crucially, adaptation, conversion, and development. Turnbull and Chambers make an important contribution here in pointing out that what ICTs make possible is as limited or as expansive as the conception of knowledge that they work to reproduce, transmit, and make available.

Incommensurability is a key issue in cross-cultural and cross-disciplinary dialogue and collaboration regarding knowing. The authors here are often explicit about this issue and its problems. Turnbull and Chambers, for example, consider how incommensurate knowledge practices, traditions, and even ontologies can be accommodated together in a new manner by ICT initiatives. In concurrence with Nafus's impetus to realign the thinking and efforts of a large technology-producing company, Turnbull and Chambers ask us to consider alternatives to merely ignoring, or otherwise colonizing and appropriating, varieties of knowledge and associated ways of being and knowing. Their contribution is to attend to how ICTs might best be able to serve this process. And in this focus, they complement the detailed and sophisticated discussion provided by Verran and Christie. Verran and Christie argue that we need to recognize that digital databases to enhance knowledge practices involve the "working together" of disparate knowledge traditions (as do Lewis, Nafus, and Diemberger and Hugh-Jones). In attempting this working together, they note that "Aboriginal knowledge workers who were developing their own private collections of digital objects were challenged by the traces of Western metaphysics apparently *inside* the software they had taken up."

**Conclusion**

As Strathern writes, in the self-proclaimed knowledge economy, information is the stuff that knowledge is constituted by. Euro-American conceptions of information and the role that it plays as the basis for action "produce a sense that it can be quantified, whether in terms of sufficiency (how much is needed) or through multiplying different compartments of it" (2006, 192). In the imaginaries of "modern high-tech capitalism" (Thrift 2002, 19), interconnectivity provides the ecological conditions in which information and knowledge can be incubated as the basis for innovation: an appropriative model of knowledge production and exchange in which particular forms of personhood, sociality, and modes of creativity prevail (Leach 2004). Alongside this appropriative emphasis when it comes to knowledge is an allied emphasis on production, circulation, and value creation through just the kinds of connectivity that ICTs enable.

This volume has demonstrated the different ways in which modes of life, trajectories of practice, ways of knowing, imaginaries, and particular values feed into and shape alternative uses and reuses of technologies in ways that have the potential to challenge this appropriative knowledge form, the epistemological assumptions, and the social values that guide most ICT production today. We consider an emphasis on ongoing and reflexive dialogue, through making and use, with people in all sorts of places and with different histories and imaginaries can only enhance the positive subversion, conversion, and ultimately development of future technologies.

## Notes

1. To insist on the global view of contemporary modernity is also to take a local, interested perspective.

2. For an example directly addressing cross-cultural knowledge exchange, see "Cross-Cultural Partnership: Template," http://newmedia.umaine.edu/stillwater/partnership/partnership_template.html.

3. We remind the reader that we have chosen the term "ICT" instead of "new media" because we intentionally blur the distinction between digital and other technologies of knowledge production and reproduction.

4. As, of course, is their terrible and proven potential for harm (Strathern).

## References

Atkins, D. E., J. S. Brown, and A. L. Hammond. 2007. A review of the Open Educational Resources (OER) movement: Achievements, challenges, and new opportunities. Report for the William and Flora Hewlett Foundation. Menlo Park, CA. http://www.hewlett.org/uploads/files/Hewlett_OER_report.pdf.

Blaser, M. 2010. *Storytelling Globalization from the Chaco and Beyond*. Durham, NC: Duke University Press.

Bowker, G. C. 2005. The knowledge economy and science and technology policy. http://www.ics.uci.edu/~gbowker/pubs.htm.

Bowker, G. C., A. Kanfer, C. Haythornthwaite, B. Bruce, N. Burbules, J. Porac, and J. Wade. 2000. Modelling distributed knowledge processes in next generation multidisciplinary alliances. http://www.ics.uci.edu/~gbowker/distribk.pdf.

Bowker, G. C., and S. Leigh Star. 1994. Knowledge and infrastructure in international information management: Problems of classification and coding. In *Information Acumen: The Understanding and Use of Knowledge in Modern Business*, ed. L. Bud-Frierman, 187–216. London: Routledge.

Bowker, G. C., and S. Leigh Starr. 1999. *Sorting Things Out: Classification and Its Consequences*. Cambridge, MA: MIT Press.

Cahill, C. 2007. Including excluded perspectives in participatory action research. *Design Studies* 28:325–340.

Cefkin, Melissa, ed. 2010. *Ethnography and the Corporate Encounter: Reflections on Research in and of Corporations*. Oxford: Berghahn Books.

Collingwood, R. G. [1940] 1998. *An Essay on Metaphysics*. Oxford: Oxford University Press.

Corsín Jiménez, A. 2011. Daribi kinship at perpendicular angles: A trompe l'oeil anthropology. *HAU Journal of Ethnographic Theory* 1 (1): 141–157.

Daston, Lorraine, and Peter Gallison. 2007. *Objectivity*. New York: Zone Books.

Deuze, M. 2006. Participation, remediation, bricolage: Considering principal components of a digital culture. *Information Society* 22 (2): 63–76.

Downey, G. L., and J. Dumit, eds. 1997. *Cyborgs and Citadels: Anthropological Interventions in Emerging Sciences and Technologies*. Santa Fe, NM: School of American Research Press.

Gell, A. 1998. *Art and Agency*. Oxford: Oxford University Press.

Hanseth, O., E. Monteiro, and M. Hatling. 1996. Developing information infrastructure standards: The tension between standardization and flexibility. *Science, Technology, and Human Values* 21 (4): 407–426.

Haraway, D. 1991. *Simians, Cyborgs, and Women: The Reinvention of Nature*. New York: Routledge.

Irani, L., and P. Dourish. 2009. Postcolonial interculturality. In *Late Breaking Papers: International Workshop on Intercultural Collaboration*, February 20–21, 2009. Stanford, CA.

Kensing, F., and J. Blomberg. 1998. Participatory design: Issues and concerns. *Computer Supported Cooperative Work: A Journal of Collaborative Computing* 7 (3–4): 167–185.

Krishna, S., and S. Madon. 2003. *The Digital Challenge: Information Technology in the Development Context*. Aldershot: Ashgate.

Latour, Bruno. 1987. *Science in Action: How to Follow Scientists and Engineers Through Society*. Cambridge, MA: Harvard University Press.

Latour, Bruno. 1993. *We Have Never Been Modern*. Cambridge, MA: Harvard University Press.

Leach, J. 2004. Modes of creativity. In *Transactions and Creations: Property Debates and the Stimulus of Melanesia*, ed. E. Hirsch and M. Strathern, 152–175. Oxford: Berghahn Books.

Leach, J. (with Dawn Nafus and Bernhard Krieger). 2009. Freedom imagined: Ethics and aesthetics in open source software design. *Ethnos* 74 (1): 51–71.

Lemonnier, Pierre, ed. 1993. *Technological Choices: Transformations in Material Cultures since the Neolithic*. London: Routledge.

Lievrouw, L. A. 2011. *Alternative and Activist New Media*. Malden, MA: Polity.

Lievrouw, L. A., and S. Livingstone, eds. 2006. *Handbook of New Media*. London: Routledge.

Neff, G., and D. Stark. 2003. Permanently beta. In *Society Online*, ed. P. Howard and S. Jones, 173–188. Thousand Oaks, CA: Sage.

Quine, W. van O. 1960. *Word and Object*. Cambridge, MA: MIT Press.

Redström, J. 2008. RE: Definitions of use. *Design Studies* 29 (4): 410–423.

Stengers, Isabelle. 2005. The cosmopolitical proposal. In *Making Things Public: Atmospheres of Democracy*, ed. Bruno Latour and Peter Weibel. Cambridge, MA: MIT Press.

Strathern, M. 1995. The nice thing about culture is that everyone has it. In *Shifting Contexts*, ed. M. Strathern, 153–176. London: Routledge.

Strathern, M. 2006. A community of critics? Thoughts on new knowledge. *Journal of the Royal Anthropological Institute* 12 (1): 191–209.

Thrift, N. 2002. "Think and act like revolutionaries": Episodes from the global triumph of management discourse. *Critical Quarterly* 44 (3): 19–26.

Winner, Langdon. 1980. Do artifacts have politics? *Daedalus* 109 (1): 121–136.

Winner, Langdon. 1989. *The Whale and the Reactor: A Search for Limits in an Age of High Technology*. Chicago: University of Chicago Press.

Woolgar, S. 2002. Five rules of virtuality. In *Virtual Society? Technology, Cyberbole, Reality*, ed. Steve Woolgar, 1–22. Oxford: Oxford University Press.

## Contributors

**Poline Bala** is Senior Lecturer at the Department of Anthropology and Sociology in the Faculty of Social Sciences at the University of Malaysia, Sarawak.

**Alan Blackwell** is Reader in Interdisciplinary Design, Computer Laboratory, University of Cambridge.

**Wade Chambers** is Instructor, Indigenous Liberal Studies, at Institute of American Indian Arts.

**Michael Christie** is Professor of Education, the Northern Institute, Charles Darwin University.

**Hildegard Diemberger** is Senior Associate in Research, Mongolia and Inner Asia Studies Unit, at the University of Cambridge.

**Stephen Hugh-Jones** is Honorary Emeritus Associate, Division of Social Anthropology, University of Cambridge.

**James Leach** is Professor of Anthropology and Future Fellow at the University of Western Australia and Director of Research at CREDO CNRS.

**Jerome Lewis** is Lecturer in Anthropology at University College London.

**Dawn Nafus** is a researcher at XIL, Intel Labs, Portland.

**Gregers Petersen** is a Danish anthropologist and activist for the OpenWrt project.

**Marilyn Strathern** is Fellow of Girton College, Cambridge.

**David Turnbull** is Honorary Research Associate/Fellow, Arts Faculty, Deakin University.

**Helen Verran** is an Australian historian and philosopher of science at the Northern Institute, Charles Darwin University.

**Laura Watts** is Associate Professor, Technologies in Practice Group, IT University, Copenhagen.

**Lee Wilson** is a Research Fellow at the School of Political Science and International Studies, University of Queensland.

# Index

Abduction, 163
Aboriginal Australian databasing, postcolonial
  and archiving, 58, 60, 72
  and ceremonies, 71–72
  and collective memory, 57, 58, 59, 61, 63, 71
  and distinction between systems and traditions, 61
  and Dreaming knowledge, 69–70
  and epistemic difference, 60, 62–63, 67, 75
  and fire knowledge, 63–68
  and intellectual property, 74
  and interaction between knowledge traditions, 57–59, 62–63, 70–72, 75
  intergenerational aspects of, 57, 74
  and local knowledge, 57, 58, 72–73, 233, 239
  and metadata, 60, 65, 68
  metaphysical aspects of, 69–70, 75
  and ontic difference, 60, 62–63, 67, 73, 75
  and performativity, 60, 65, 68, 72, 73
  political aspects of, 60
  social aspects of, 71–72
  and technoscientific traditions, 57, 61, 62, 63–71
  and working imaginaries, 58
Actor network theory, 158
Adoption, technological, 5, 29, 34, 40, 79, 83, 85, 89, 100, 106, 113, 116–118, 122, 195, 196, 197, 212–216, 220, 233, 237, 239, 241
AIDS, 158, 159
Alaska, indigenous knowledge in, 166, 170
Amazonia
  books as ritual objects in, 90–100
  human impact on ecology of, 157–158
  indigenous knowledge in, 156–157
Amster, Mathew H., 109
Anarchism, 49
Anderson, Ken, 188, 211
Annan, Kofi, 30
Antennae, in wireless network devices, 41, 42, 45, 47–48
Anthropology
  in the Amazon, 91, 156
  and book artifacts, 83, 91
  and colonialism, 183
  corporate, 186, 188, 190, 193, 194, 201–202, 207, 208–209, 210, 219–220
  as critical reflection, 2
  and design, 187, 188, 194, 198, 201
  economic, 206, 211, 216, 220
  and egalitarian societies, 137
  and knowledge production, 240
  and market relations, 211, 216, 220
  and nonhuman agency, 84
  and process-based knowledge, 2
  and quantitative knowledge, 218
  and scale, 211

Aporta, Claudio, 166
Apple corporation, 193
Appropriation, 1
　and consumer agency, 201, 203
　and Freifunk wireless network (Berlin), 39, 40
Aramis transit system, in Paris, 227
Archives and archiving, 58, 60, 72, 80, 89, 98, 153, 168, 172, 195, 235
Artworks, 3, 84
Attfield, Judy, 52
Australia
　didgeridoos made in, 173
　postcolonial databasing in, 57–75

Bacon Hales, Peter, 11
Bahrin, Tengku S., 109
Bala, Poline, 232, 237, 239, 240
Balée, William, 157
Barrett, Tim, 85
Baskin, Ken, 166
Bell, Genevieve, 188
Bellah, R. N., 110
Berlin
　economic relations in, 42–43, 48
　Freifunk wireless network in, 39, 42–43, 46
　leftist political activity in, 49
Bertalanffy, Ludwig von, 160
Bijker, Wiebe E., 52
Biogas industry, 23, 25, 26, 31
Blackwell, Alan, 4
Bodily processes, 226, 228
Books, impact of digital technology on, 79, 90, 100. *See also* Amazonia, books as ritual objects in; Buddhism, books as sacred relics in
Borneo. *See* Kelabit people, in Central Borneo
Boundary object, 192, 198, 210, 217, 219
Bowker, G. C., 236
Brazil
　Amazonian Indians in, 91, 156, 157
　as emerging market, 210, 211, 213
Bricolage, 12–13, 46, 234
Bride wealth, 7, 226, 227
Bridging the Global Divides (BGDD), 195–197
Brosius, P., 110
Buddhism
　books as sacred relics in, 79, 84–86
　Cultural Revolution's impact on, 81, 91, 98
　and digital books, 79–83, 86–91, 98, 99–100
　and Tibetan-Mongolian Rare Books and Manuscripts (TMRBM) Project, 79–83, 90

Cajete, Gregory, 170
Callon, Michel, 202, 206
Cameroon, forest management in, 128, 147–149
Campbell, Donald, 162
Capitalism, 49, 52, 54, 153, 219, 240, 241
Chambers, Wade, 231, 236, 239, 240, 241
Chandra, M., 110
Chaos Computrer Club (CCC), 39
Chile, wine producers in, 195
China
　books and printing in, 85
　Cultural Revolution in, 81, 91, 98
　as emerging market, 205, 209–210, 212, 216
　Tibet ruled by, 80–81, 99
Christianity, 70, 108, 117
Christie, Michael, 232, 233, 235, 239, 240, 241
Clammer, J., 106
Clans, 225–226
　in Amazonian Brazil, 91
　in Congo Basin, 135, 140, 141
　in Tibet, 87

# Index

Clastres, Pierre, 49
Climate change, 157–158
Collective memory, 57, 58, 59, 61, 63, 71
Colonialism, 183
  in Australia, 58
  in Central Africa, 129, 131, 132
  in South America, 156
  in Tibet, 81, 83
Complex adaptive systems, 158, 159–167, 171–173
Computer science, 183, 185, 188. *See also* Digitial technology; ICT
Congo. *See* Logging industry, in the Congo Basin; Pygmies in the Congo Basin
Conservationism, 97, 129–131, 135–136, 147, 150, 239
Consumer agency, 201–206
  and appropriation, 201, 203
  prediction of, 201–206, 210–211, 213, 215, 216, 218–220
  and reuse, 203–205
  and user-centered design, 201, 203–205
Consumerism, 21, 184
Corsín Jiménez, A., 234
Crabtree, A. T., 192
Creativity
  and complex adaptive systems, 164
  and improvisation, 13–14
  and innovation, 13, 35, 40
  process-based, 3
  relational, 33, 34
  social aspects of, 6, 11, 14, 43, 241
  and subversion, 5, 40
Cross-cultural exchange, 2, 3, 6, 14, 84, 100, 185, 241
Cruikshank, Julie, 170
Cussins, Adrian, 174n12

de Castro, Viveiros, 156
decentralization, 23, 29
de Certeau, Michel, 51, 53, 122
Deleuze, Gilles, 163
Denevan, William, 157
Design, technological and industrial. *See also* Software design
  and consumer agency, 201–204, 219
  and design research, 190–198
  and ethnographic engagement, 3–5
  and ethnomethodology, 188–190
  and Freifunk wireless network (Berlin), 40
  global adoption of, 194–197
  in Orkney, 32, 33, 34
  political aspects of, 1–2, 40, 237, 238
  and prototypes, 191–192, 194–195
  and subversion, 5, 197–198, 219, 237
  user-centered, 201, 203, 204, 205, 218, 238
Design science, anthropology as, 187
Deuze, Mark, 234
Development
  socioeconomic, 105, 110–113, 115–123, 237
  technological, 30, 105–106, 113, 237
Dewey, John, 163
Diamond, Jared, 158
Didgeridoos, 173
Diemberger, Hildegard, 81, 82, 232, 233, 236, 240, 241
Digital media, 89–90, 100, 237
Digital objects
  and Aboriginal Australian databasing, 57, 59–60, 63, 65, 68, 70, 71, 73, 75, 233, 241
  Buddhist sacred books as, 83, 88
  and Western knowledge traditions, 70
Digital technology
  and Aboriginal Australian databasing, 57–58, 59–60, 65, 71–72, 73
  and Amazonian Indian book-objects, 98–99
  and Buddhist sacred books, 79–83, 86–91, 98, 99–100

Digital technology (cont.)
  and design prototypes, 192
  and ethnographic description, 194
  and Kelabit people in Central Borneo, 105, 106, 107, 113–123
  printed book format impacted by, 79, 90, 100
Digitization, 2, 232
  and Aboriginal Australian databasing, 58, 59, 65, 71, 72, 73–74
  and Buddhist sacred books, 80, 82, 83, 87–88, 89
Distributed knowledge, 165, 236
Distributed personhood, 84
Distributed systems, 171, 173
Diversity, social, 224–225
Dogs' teeth, in Papua New Guinea, 6–10, 234
Downward causation, 162
D-Space, 83
Dussel, Enrique, 156

Easter Island, 158
eBario development project, 105, 106, 107, 113–123
Economic relations
  and anthropology, 206, 211, 216, 220
  in Berlin, 42–43, 48
  and Callon's work, 202, 206–210, 215, 219, 220
  in Central Borneo, 107, 111–112, 120
  and GDP (gross domestic product), 212
  and knowledge economy, 153, 231, 235, 241
  in Malaysia, 107, 110–112
  and Miller's work, 201, 206–207, 215, 220
  neoclassical theory of, 202, 206, 207, 209, 212, 215, 220
  and performativity, 207, 209, 212, 220
  and predictability, 201–206, 210–211, 213, 215, 216, 218–220

Education, in Central Africa, 129
Egalitarianism, 128, 129–130, 137–139
El Dorado, mythical city of, 157
Embeddedness, 2, 3, 31, 32, 34, 35, 43, 45, 58, 224, 232, 236, 239
  and industrial design, 202, 206, 207, 209, 212, 215, 217, 218, 219
Emergence, in complex systems, 1, 2, 12, 14, 159, 160, 162, 164, 165, 167, 168, 169, 174–175nn14–15
Emerging markets, 203, 205, 210–212, 216–217
Engelbart, Douglas, 188
Epistemology
  and complex adaptive systems, 161, 164, 165
  and indigenous knowledge, 62, 64, 75, 157, 158, 159, 167, 171
  relational, 2, 156, 158
  and scientific knowledge, 154
  and technological design, 1, 3
Erickson, Clark, 157
Ethnic identity, in Malaysia, 110–111, 117, 121, 122
Ethnography
  corporate, 193–194, 210–212
  and design, 188–190, 197
  and ethnographic engagement, 3–5
  and knowledge production, 240
  and technological improvisation, 32, 41–42, 79
Ethnomethodology, 189–190
Exaptation, 12–13
Exogamy, 225–226
Extreme weather, in Orkney, 21, 31

Facebook, 184
Fair Tracing project, 195
Family relations, suburban, 12, 13, 14. *See also* Kinship
Feedback loops
  and complex adaptive systems, 162, 164, 167, 169

# Index

and industrial design, 201, 209
and stigmergy, 172, 175n20
Feenberg, Andrew, 52
Fiber optics, 20, 43
Ford, Henry, 11
Forest management
  in Cameroon, 128, 147–149
  in the Congo Basin, 127–145, 147, 149–151
Frames, in socioeconomic theory, 202, 206, 207, 208, 215, 217
Free software, 2, 39, 41, 43, 234
Freifunk wireless network (Berlin)
  definition of, 39
  history of, 39, 40–44, 46–47
  as mesh network, 39, 41–43, 45–52, 54
  political aspects of, 39–40, 43, 49–54
  social aspects of, 39–40, 48–54
  and subversion, 39, 40, 50–53
  technological aspects of, 39, 42, 44–46, 47–48
Future-making, in Orkney, 19–36

Garfinkel, Harold, 189
GDP (gross domestic product), 212
Gell, Alfred, 84, 100n3
Gender identity, in Papua New Guinea, 7, 10
Geurts, Kathryn, 171
Ghana, indigenous knowledge in, 171
GIS (geographic information systems), 127, 133, 135, 149
Globalization, 2, 100, 122, 227
God trick, Haraway's concept of, 215, 232
Goffman, Erving, 207
Goldenfeld, Nigel, 163
Gonzales, Patrick, 166
Goody, Jack, 86
Google, 193
GPS (Global Positioning System), 10, 121, 127, 133, 135, 142–143, 196
Graeber, David, 51

Grassé, Pierre-Paul, 171
Guattari, Félix, 163

Hacking, 32, 39, 46, 49, 203
Haraway, Donna, 174n15, 215, 219, 232
Harris, M. A., 190
Harrisson, Tom, 107, 116
Heylighen, Francis, 171–172
Hierarchy
  and Amazonian Indian society, 93
  and Malaysian society, 110–111
  and politics of wireless networks, 44, 47, 49, 52
Hilley, J., 110
Holland, John, 166
Horizontalism, political, 43, 46, 49, 50, 54
Howard, Phil, 212
Hugh-Jones, Stephen, 82, 98, 232, 233, 236, 240, 241

IBM corporation, 193
IBM/World Bank e-readiness index, 212
ICT (information and communications technology). *See also* Freifunk wireless network (Berlin)
  and Australian Aborigines, 57, 58, 59, 61, 63, 75
  and consumer agency, 201
  and ICT [for] Development (ICT4D), 195–196
  immateriality of, 184
  and indigenous knowledge, 57, 58, 59, 61, 63, 75, 158, 171, 231, 240
  and Kelabit people in Central Borneo, 105, 112, 113–123, 239
  and knowledge production, 1–2, 236, 239, 240, 241
  in London, 20
  in Orkney, 20, 30, 32, 33, 36
  and prototypes, 195, 196–197
  social aspects of, 5–6, 10, 183, 231, 232, 235

ICT (cont.)
  and social/technical development, 237–238
  and software design, 184, 185
  and subversion, 5
  and Technology Distribution Index (TDI), 213
Illiteracy. *See* Nonliteracy
Imaginaries, 5, 6, 20, 58, 71, 211, 233, 235, 237, 241, 242
Immateriality, 184, 185, 194
Improvisation
  cultural, 237
  distinguished from innovation, 13
  technological, 32, 41–42, 79
Incommensurability, 64–65, 156, 159, 161, 168, 174, 211, 239, 241
India
  indigenous knowledge in, 155
  radio in, 195
Indigenous knowledge, 2–3, 57–75, 154–159, 168–173, 231, 240
Information technology. *See* ICT (information and communications technology)
Infrastructure
  in Berlin, 42–43, 53
  in Orkney, 21–23, 30–36
Ingold, Tim, 12–13, 166, 219
Innovation
  and Aboriginal Australian databasing, 73
  distinguished from improvisation, 13, 14
  landscape as influence on, 19, 34, 35
  in Orkney, 19, 25–27, 31, 233
Intel Corporation, 188, 193, 201, 204, 205–210, 212–213, 215–216, 218, 232
Intellectual property, 1, 74, 224
Interests, social, 224–227, 228
Internet. *See also* Wireless networks
  and Aboriginal Australian databasing, 74
  and Berlin Freifunk, 39, 43–44, 46, 48, 50, 54
  and Cameroonian forest management, 147
  and indigenous knowledge, 74, 154–155, 168–169
  and Kelabit people in Central Borneo, 105, 106, 107, 113–114, 116–120, 122, 123
  and knowledge production, 2
  and proliferation of knowledge, 153
Intertextuality, 166
Inuit people, 166
Islam, 70, 110

Jacoby, Sally, 166
Jordan, Brigitte, 189

Kadanoff, Leo P., 163
Kayapo people, in South America, 97, 98–99, 100, 101n13
Kelabit people, in Central Borneo
  Christianity adopted by, 108, 117
  and cultural borrowing, 108, 116–117
  development sought by, 106, 109, 111–112, 115–123
  and *doo*-ness (social goodness), 106–110, 111–113, 114, 115, 118, 119, 122
  and eBario project, 105, 106, 107, 113–123
  ICT adopted by, 105, 112, 113–123, 239
  and *iyuk* (social status), 106–110, 111–113, 114–115, 116, 117, 122
  language of, 120, 123n3
  logging industry's impact on, 120, 121
  and Malaysian state, 109–113, 117, 118–119, 121, 122
  and migration, 109–110, 116–117, 119
  and performativity, 108–109

# Index

progress sought by, 106–107, 109, 111–113, 114–117, 118–119, 122
Kidwell, Clara Sue, 170
King, V. T., 110
Kinship
  in Amazonian Brazil, 93
  in Central Borneo, 119
  in Russia, 211, 212
  in Tibet, 87
Knowledge
  assemblage of, 153, 156, 159–160, 163–165, 167
  and capitalism, 153
  and complex adaptive systems, 158, 159–167, 171–173
  distributed, 165, 236
  and ethnographic engagement, 3–5
  indigenous, 2–3, 57–75, 154–159, 168–173, 231, 240
  on Internet, 153, 154–155, 168–169
  and knowledge commons, 153, 154, 168
  and knowledge economy, 153, 231, 235, 241
  and knowledge production, 1–2
  and knowledge society, 105, 113, 122
  and linguistic diversity, 155
  local, 57, 58, 72–73, 154, 155, 157, 165, 231, 233, 239
  and metaphysics, 69–70
  and multiplicity, 154, 156–157, 159–165, 167, 168, 237
  performative, 164, 165, 166
  process-based, 2, 3
  proliferation of, 153–154
  relational, 2, 3, 156, 158
  situated, 3, 240
  social aspects of, 163–165, 236
  tacit, 153, 165, 171
  technoscientific, 3, 57, 61, 62, 63–71, 154, 159, 197, 240. *See also* Epistemology

Landscape
  anthropogenic, 157
  and ICT, 20, 30
  innovation affected by, 19, 34, 35
  and technological imaginary, 20, 233
Larsen, A. K., 110
Latour, Bruno, 156–157, 215, 227, 228
Law, John, 52
Lee, B. T., 109
Lemmonier, Pierre, 227
Levebvre, Henri, 51
Lévi-Strauss, Claude, 12–13
Levitt, William, 11, 14
Levittown, New York, suburban design in, 10–12, 13, 14, 234
Lewis, Jerome, 232, 236, 237, 239, 240, 241
Lievrouw, L. A., 231, 234
Linguistic diversity, 155
Linux, 39, 41, 46, 47, 220
Livingstone, S., 231, 234
Local knowledge, 57, 58, 72–73, 154, 155, 157, 165, 231, 233, 239
Logging industry
  in Cameroon, 147–149
  in the Congo Basin, 127, 130–137, 139–142, 145, 147
  in Malaysia, 120, 121
London
  mobile telecom industry in, 20
  wireless networks in, 43
Long, Norman, 122
Lovelock, James, 158

Malaysia. *See also* Kelabit people, in Central Borneo
  ethnic status in, 110–111, 117, 121, 122
  as knowledge-based society, 105, 113, 122
  logging industry in, 120
  socioeconomic development in, 110–113
Maori people, 169–170

Maps
  Cameroonian villagers' use of use of, 147–149
  logging industry's use of, 127, 132–135
  Pygmies' use of, 128, 140–145, 147, 149–151, 239
  software developed for, 128, 142–145, 147, 148–149, 150
  and technology adoption, 214, 232
Market relations, 42–43, 201, 203–213, 215–218, 220, 234, 237
Massey, Doreen, 161
Maturana, Humberto, 162, 163
Media. *See* Digital media; ICT (information and communications technology); Internet; Mobile telecommunications; Printing, in Buddhist societies; Radio
Mesh networks, 39, 41–43, 45–52, 54
Metaphysics, 69–70, 75, 161, 162, 240, 241
Microsoft corporation, 193
Migration, in Central Borneo, 109–110, 119
Millen, David R., 4
Miller, Daniel, 114, 201, 206, 207, 215, 220
Mobile telecommunications, 20, 34, 195
Modernity, 13, 98, 99, 105, 116, 119, 156–157, 170, 172, 235, 242n1
Monbiot, George, 158
Mongolian books and manuscripts. *See* Tibetan-Mongolian Rare Books and Manuscripts (TMRBM) Project
Moore's Law, 20
Multinaturalism, 156
Multiplicity, in knowledge systems, 154, 156–157, 159–165, 167, 168
Mumford, Lewis, 11

Nafus, Dawn, 194, 232, 233, 234, 239, 241
Narrative, knowledge systems based on, 154, 159, 162, 163, 165, 166–167, 170

Native Americans, 169, 170, 174n13
Neff, G., 232
Negri, Antonio, 53
Neoclassical economics, 202, 206, 207, 209, 212, 215, 220
Neolithic artifacts, 20, 27, 31
Networks. *See* Mesh networks; Wireless networks
Neutrality, technological, 1, 10, 53, 204, 209, 233
Nonliteracy
  and Amazonian Indians, 90–91, 94
  and Cameroonian villagers, 147–148
  and Congo Pygmies, 129–130
  and Tibetan Buddhists, 88, 90–91

Ochs, Elinor, 166
Ontology
  and complex adaptive systems, 160–165, 167, 175n15
  and indigenous knowledge, 60, 62, 65–69, 73, 154, 156–157, 159, 167, 169, 170, 174n14
Open-source software, 35, 54, 220, 234
OpenWrt.org, 42, 47
Orellana, Francisco de, 157
Orkney Islands, 19–36, 233
Ortner, S. B., 106
Overflow, social, 202, 207, 208, 215, 217, 218

Papua New Guinea, 3, 6–10, 226, 227
PARC. *See* Xerox
Paris, Aramis transit system in, 227
Performativity
  and Aboriginal Australian databasing, 60, 65, 68, 72, 73
  and economic relations, 207, 209, 212, 220
  and Kelabit people in Central Borneo, 108–109
  and knowledge systems, 164, 165, 166, 240–241

# Index

Petersen, Gregers, 232, 234, 235
Phuntsho, Karma, 82
Pickering, Andrew, 175n15
Plato, 70, 153
Polanyi, Michael, 153
Political relations
   and Aboriginal Australian databasing, 60
   and Amazonian Indian ritual objects, 93, 98
   and egalitarianism, 128, 129–130, 137–139
   and Freifunk wireless network (Berlin), 39–40, 43, 49–54
   and hierarchy, 44, 47, 49, 52, 93, 110–112
   and horizontalism, 43, 46, 49, 50, 54
   and Kayapo (South American) resistance actions, 98–99
   and Kelabit people in Central Borneo, 111, 112, 239
   and network protocols, 44–45
   and Pygmies in the Congo Basin, 128
   and technological design, 1–2, 40, 237, 238
   and technological subversion, 1, 50–53
POM beverage, 203
Postcolonial databasing. *See* Aboriginal Australian databasing, postcolonial
Predictability of economic behavior, 201–206, 210–211, 213, 215, 216, 218–220
Printing, in Buddhist societies, 79, 85, 87
Privatization, 42
Prototypes, technological, 191–192, 194–195, 196–197
Pygmies in the Congo Basin
   and anti-poaching measures, 131, 149
   avoidance practiced by, 127, 131
   egalitarianism practiced by, 128, 129–130, 137–139
   excluded from policy decisions, 127, 135, 149
   as forest dwellers, 127, 130–132, 135
   maps used by, 128, 140–145, 147, 149–151, 239
   and nonliteracy, 129–130
   and relations with logging industry, 130–132, 130–137, 139–142, 145, 147
   software designed for, 128

Race relations, 14n3
Radio
   in Central Africa, 147
   in India, 195
Reconfiguration, 5–7, 10, 12, 228, 234, 235
   and Aboriginal Australian databasing, 57–58
   of dogs' teeth in Papua New Guinea, 7, 10
   and Freifunk wireless network (Berlin), 47
   in Orkney, 31–35
Redström, J., 238
Reflexivity, 4, 11, 58, 59, 163, 164, 186, 191, 192, 197, 228, 229n6, 231, 236, 239, 240, 242
Religion, 69–70. *See also* Buddhism; Christianity; Islam
Renewable energy, 19, 31, 35–36. *See also* Wind energy
Repurposing, 5, 7, 10, 12–14, 61, 203–205, 231, 234, 236
Reuse, 5, 7, 12, 13, 32, 51, 203–205, 218, 234, 242
Reynolds, Malvina, 11
Rinpoche, Chokyi Nyima, 89
Rinpoche, Zenkar, 81, 82
Ritual objects, Amazonian Indians' books as, 90–100
Robertson, A. F., 110
Royal Society of London, 224, 227
Rulifson, Jeff, 188
Russia, 205, 209–210, 211, 212

Salesian missionaries, in Brazil, 91, 96, 97
Salvador, Tony, 188
Scale
 in anthropology, 211
 in complex systems, 161–162
 in market relations, 210–211
Schools. *See* Education
Scientific knowledge, 3, 57, 61, 62, 63–71, 154, 159, 197, 240
Scott, J. C., 110
Seattle, wireless networks in, 43
Self-organization, 160, 162
Self-sufficiency, in Orkney, 23–25, 31, 32, 33
Sennet, R., 12
Simon, Herbert, 187
Situated futures, 19, 20, 31–32, 34–35
Situated knowledge, 3, 240
Situated technology, 32, 34
Slater, D., 114, 207
Smith, Gene, 81, 82
Smith, Linda Tuhiwai, 169–170
Smolin, Lee, 162
Social justice, 39–40
Social relations
 and Aboriginal Australian ceremonies, 71–72
 and Amazonian Indian patrimony, 93
 and Buddhist sacred books, 87
 and Congo Pygmies, 137–139
 and creativity, 6, 11, 14, 43, 241
 and Freifunk wireless network (Berlin), 39–40, 48–54
 and ICT, 5–6, 10, 231, 232, 235, 236, 237
 interests in, 224–225, 228
 and Kelabit people in Central Borneo, 106–110, 112
 and knowledge production, 163–165, 236
 in Orkney, 32–33
 in Papua New Guinea, 6–7, 10, 226, 227
 and technology, 1, 4, 39–40, 41, 105–106, 113, 128–129
Social science, 183–184, 185, 197. *See also* Anthropology; Ethnography
Software design
 and anthropology, 187, 188, 194, 198, 201
 and design research, 190–193
 engineering and craft conflated in, 184–185
 and ethnography, 193–194
 and global adoption, 194–197
 and immateriality, 194
 and knowledge production, 2
 and mapping for nonliterate users, 128–129, 142–145, 147, 148–149, 150
 and prototypes, 192, 194–195
 and requirements analysis, 186
 structure as defining feature of, 184, 185
South Africa, indigenous knowledge in, 155, 158
South America, indigenous knowledge in, 155–156
Spatiality, in knowledge systems, 161, 163–168
Stanford Research Institute, 188
Stark, D., 232
Stigmergy, 164, 171–172, 175n20
Storied spaces, 166, 167, 168, 169, 171, 173
StoryBank project, 195
Strathern, Mariyln, 231, 232, 234
Suburbia, residential design in, 10–12
Subversion, 1, 5, 234, 237
 and design research, 197–198, 219
 and digital impact on ritual practice, 99
 and Freifunk wireless network (Berlin), 39, 40, 50–53
 and map of technology adoption, 232
Suchman, Lucy, 188, 189, 193
Sustainability, in Orkney, 31, 32, 33, 35

# Index

Tagging, in complex adaptive systems, 166, 172
TASSIT Project, 168–171
Technology. *See also* Adoption, technological; Design, technological and industrial; Digital technology; ICT (information and communications technology)
  development of, 30, 105–106, 113, 237
  and Freifunk wireless network (Berlin), 39, 42, 44–46, 47–48
  of hunting, in Papua New Guinea, 6–7
  and knowledge production, 1–2, 240
  political aspects of, 1, 39–40, 50–54, 234
  purported neutrality of, 1, 10, 53, 204, 209, 233
  situated, 32
  social aspects of, 1, 4, 39–40, 41, 105–106, 113
Technology Distribution Index (TDI), 212–219
Temporality, in knowledge systems, 162, 163, 164, 165, 166, 168, 171
Termite mounds, 171–173
Tibetan-Mongolian Rare Books and Manuscripts (TMRBM) Project, 79–83, 90
Tinker, George, 170
Trails, knowledge systems as, 156, 165–169, 171–172, 174n12, 175n18
Tukanoan language and culture, 82, 90–91, 93–100
Turnbull, David, 231, 236, 239, 240, 241
Turner, J. Scott, 175n20
Turner, V., 108–109
Twitter, 184

UNESCO, 159
United Kingdom, design research in, 195–196
User-centered design, 201, 203, 204, 205, 218, 238

Varela, Francisco, 162, 163
VCRs, 79, 98–99
Verran, Helen, 232, 233, 235, 239, 240, 241
VSATs, 105, 112

Waddell, Austin, 81, 83
Walsham, Geoff, 196
Ward, Colin, 51
Watts, Laura, 233, 235
Web. *See* Internet
Wind energy, 19, 24, 25, 27, 28, 31
Wireless networks
  and Freifunk (Berlin), 39–54
  structure of, 44–46
Woolgar, S., 232
World Bank, 212
Wynn, Eleanor, 188, 193

Xerox EuroPARC, 189, 192
Xerox Palo Alto Research Center (PARC), 188, 193

Yaka people. *See* Pygmies in the Congo Basin